GREEN SEDUCTION

EG08118860D
G7
THIS NOTE IS LEGAL TENDER
FOR ALL DEBTS, PUBLIC AND PRIVATE

GREEN SEDUCTION

MONEY, BUSINESS, AND THE ENVIRONMENT

BILL STREEVER

UNIVERSITY PRESS OF MISSISSIPPI / JACKSON

www.upress.state.ms.us

The University Press of Mississippi is a member of the
Association of American University Presses.

Manufactured in the United States of America

First edition 2007
∞

Portions of Chapters 2 and 8 were published earlier in a somewhat
different form in *Environmental Design and Construction* magazine.

Library of Congress Cataloging-in-Publication Data
Streever, Bill
Green seduction : money, business, and the environment / Bill Streever. — 1st ed.
p. cm.
Includes index.
ISBN-13: 978-1-57806-920-0 (alk. paper)
ISBN-10: 1-57806-920-3 (alk. paper)
1. Environmentalism—History. 2. Environmental protection—Economic aspects. I. Title.
GE195.S774 2007
333.7209—dc22 2006011228

British Library Cataloging-in-Publication Data available

FOR ISHMAEL STREEVER AND LUCY SLEVIN,
WHOSE FUTURE, WITH ALL OF OURS, DEPENDS ON
RESPONSIBLE ENVIRONMENTAL STEWARDSHIP

CONTENTS

GREEN SEDUCTION

CHAPTER ONE

A HALF SUCCESS

As a younger man, Grant Ferrier dreamed of changing the world. With a Berkeley engineering degree, he hoped to pursue graduate research on solar energy. At the time—this was in the early eighties—he believed that America would look past cost barriers and invest heavily in an alternative to fossil fuels. Although not driven by the pursuit of personal wealth, he foresaw a bright future. He foresaw a career rewiring the nation, pulling the fuse on coal-fired power plants, extinguishing the flames of petroleum fires, and, in the end, leaving the world a better place.

None of this worked out. Instead, Grant manages, writes for, edits, publishes, and distributes a news bulletin for the environmental industry. The *Environmental Business Journal*, which Grant sometimes describes as "the *Wall Street Journal* of the environmental industry," is printed on eight-and-a-half-by-eleven-inch paper. Under the masthead, each issue explains the rationale behind the journal with a simple phrase, presented in italics: "*Strategic Information for a Changing Industry.*" He prints six issues a year, each giving twenty pages or so of articles with headlines like "Renewables Markets Look for Boosts from

Incentives in Energy Bill," and "US Filter Aims for Greater Global Presence following Acquisition by Siemens," and "Uncertainty in Air Policy still Confounds Pollution Control Decisions." Dry stuff, the kind of stuff that Grant himself calls "strategic intelligence," but the *Environmental Business Journal* provides insights that you will find nowhere else. If you are accountable for part of the nation's $230-billion-a-year environmental industry, if you run one of the nation's thirty thousand environmental companies, or if you manage some of the almost one and a half million environmental industry employees, this is important stuff, the kind of strategic intelligence that can help you bring together environmental aspirations and the money to pay for them.

We talk in the San Diego office from which Grant runs the *Environmental Business Journal.* Though he is now in his forties, the Berkeley culture remains. In the middle of the business day he talks to me in stockinged feet, with a three-day growth of beard, while a cat—a Burmese named Sophie, more black than gray—strolls across the conference room table, sniffs his tea, steps deftly onto his lap, and settles in to have her head rubbed. The office itself has the feel of college: a certain amount of clutter, a *National Geographic* poster of "This Dynamic Planet," worn carpets, and a half-eaten cake sitting on top of a filing cabinet.

"Most of the funding for alternative energy was removed by the Reagan administration," Grant says. "Reagan's Department of Energy yanked the funding. I was totally idealistic. I thought I could solve the world's problems by making solar power. The lesson from the Reagan administration was simple. The lesson said, 'Our economy values oil and we will use all the oil we can.' The economy is market driven. The lesson said, 'It's cheaper today, so go with it.'"

He pauses for a moment, sips from his mug of tea, and leans back in his chair. "That," he adds, "left a stamp on me."

Although he does not mention it, the Reagan administration underscored its disdain for solar energy by removing, at no small cost, the solar water heaters that Jimmy Carter had installed on the White House roof. The logic, to Reagan, was simple. It made sense to encourage consumption. In Reagan's view, conservation, or for that matter any restraint in the consumption of natural resources, was bad for morale, bad for the economy, and bad for America. Coal could be used to generate electricity for under four cents per kilowatt hour, while wind energy at that time cost almost nine cents per kilowatt hour and solar cost twenty-five cents per kilowatt hour. Reagan's vision was straightforward: Promote cheap power and let Americans consume at will.

"What about the uncounted externalities of pollution and waste generation?" Grant asks. "I'd looked at that from an academic standpoint, but they hammered it into me that these things were not considered in the reality of the free market. It wasn't the government's job to meddle in the energy market. So you had the conservative business oriented approach. Let the market figure it out."

Air was free. Someone paid for uncounted externalities like soot and acid rain and smog and asthma and even premature deaths, but these costs never showed up on an electricity bill. On the balance sheets, coal was a good deal. Oil was a good deal. Natural gas was not too bad. Wind and solar were impractical ideas that only Berkeley crackpots and their graduate students could love.

"Since I couldn't figure out how to make money by solving environmental problems myself," Grant says, "I thought, well, who is doing this? Who's making money and why? So I started writing about it. The first issue of the *Environmental Business Journal* came out in October of 1988. It was completely funded by me and one of my original partners. We started the business with twenty thousand dollars. We sent out three mailings of five thousand copies. We didn't pay ourselves salaries for the first six months."

Thirty-five years ago, not long before Grant started the *Environmental Business Journal*, rivers burned. We dumped untreated sewage into our lakes and estuaries. In certain neighborhoods, a toxic mix of petrochemicals and water seeped from the ground in grotesque rivulets, leaving stinking puddles and pools beneath barbecue grills and under shiny red-and-green swing sets. We built highways and dams and irrigation systems without thinking about environmental impacts. We logged forests and mined virgin earth without pause. Our smokestacks bled carcinogens and acids. On clotheslines, shirts and pants and bed sheets filtered soot from the breeze. We filled swamps and marshes—not yet using the word "wetlands"—and then wondered why our streets flooded when it rained. We understood little about our world. Ecologists were kooks.

The Cold War and the Vietnam War and jobs and kids overwhelmed us. Frisbees and three black-and-white television channels distracted us. Eventually, though, we could no longer ignore pools of goop under swing sets and soot on our windowsills. Thirty-five years after the fact, it seems as though we awoke abruptly, startled, murmuring, "Rivers aren't supposed to burn, are they?"

Some resigned themselves to the mess, writing it off as the cost of progress, a business overhead. Others grew angry and demanded to know who did this to their world. Senator

Gaylord Nelson's first Earth Day in 1970 mobilized twenty million people, something like one in ten Americans. Public sentiment swung toward environmentalism. The democratic process went to work. The nation gave birth to the Environmental Protection Agency. New laws were passed in rapid succession: the National Environmental Policy Act, the Clean Water Act, the Safe Drinking Water Act, the Clean Air Act, the Endangered Species Act, the Marine Mammal Protection Act, the Resource Conservation and Recovery Act, and the Comprehensive Environmental Response, Compensation, and Liability Act.

Our world changed, and change brought opportunity. Government jobs sprung up, not just with federal agencies but with cities, counties, states, tribes, boroughs, and parishes. In the nonprofit sector, organizations rose up to mobilize public opinion, to look over the shoulders of government agencies, to buy land, and to offer friendly and not-so-friendly advice. In medium-sized cities, the Yellow Pages grew with advertisements for environmental consultants. Within mainstream companies, environmental departments grew from one guy with a tie—a shanghaied industrial hygienist or safety officer tasked to deal with these new laws—into whole departments, eventually spawning a new kind of corporate vice president. Our world changed, and almost overnight the environmental movement gave people the opportunity of doing well while doing good.

But our world continues to change. Even with substantial progress behind us, even with victory after victory leading to a cleaner world, there is no celebration. Even with rivers extinguished, with laws in place, with environmental nonprofit organizations that once operated on dreams and hope now bringing in and spending more than $6 billion a year, with environmental companies offering what thirty years ago was an unimagined range of services—even with all of this, a new sense of dissatisfaction has emerged.

"This is a half success story," Grant says. "We've made progress, but we're wildly unsustainable." He leans back in his chair and scratches Sophie's head. "Continuing improvement requires monetization of economic incentives. The next step forward is economic drivers for environmental performance and sustainability."

Grant and the *Environmental Business Journal*, it has been said, manufactured an industry. There may be some truth to this, but it would be closer to the mark to say that they cobbled together something coherent from what was and what remains a very diverse group of companies sharing nothing more in common than a desire to earn money by improving the environment. These are companies that have been seduced by the idea that saving the environment not only can be but must be a moneymaking proposition.

"The environmental industry was sanitation and drinking water," Grant says. "Then 1970 hit. The Environmental Protection Agency, the National Environmental Policy Act, the Clean Air Act. One after another. Anyone with a business card got more work than they could handle. Through 1991, the industry grew in double digits, both in terms of the number of companies and in terms of the size of companies."

Grant, in manufacturing an industry, created what he calls the EBJ Stock Index—the *Environmental Business Journal*'s answer to the Dow Jones, the S&P, and the NASDAQ. The EBJ Stock Index breaks the industry into sectors. But forget about activists chained to trees and Greenpeace skiffs heading off fishing trawlers on the high seas. The EBJ Stock Index sectors have all the earmarks of suit-and-tie mediocrity: the Solid Waste Management sector, the Consulting and Engineering sector, the Instrument Manufacturing sector, the Air Pollution Control sector, the Remediation Services sector. And the journal itself, though not devoid of phrases like "sustainability" and "renewables," is peppered with the jargon of business: "lingering market softness," "margins and balance sheets," "price erosion," "equity financing," "customer focus." When the government wants to know about environmental spending in the private sector, it comes to Grant. There are dollar facts, tabulated in the *Environmental Business Journal*, that are tracked nowhere else in the world. Examples: Wastewater treatment brought in $31 billion in 2003, up by a factor of three since 1980; environmental consulting brought in $20 billion in 2003, up by a factor of fifteen since 1980; instruments and information systems for tracking pollutants brought in $4 billion in 2003, up by a factor of ten since 1980. Add the sectors together—Grant tracks fourteen of them—to get a $230-billion-a-year industry, up by a factor of almost four since 1980. And this includes only commercial environmental work, leaving out the nonprofits and the government agencies and the environmental departments embedded within manufacturing and mining and timber companies.

Where other environmental journals have photographs of polar bears and whales and swamps, the *Environmental Business Journal* has bar graphs tracking the value of the industry sectors over time. Until the late 1990s, the bars shot upwards, but since then they have leveled off. The industry has peaked. Perhaps worse, bars tracking the value of the industry as a percentage of the nation's economy shot upward through 1995, then dipped.

In graduate school, Grant had counted on government subsidies to support what he saw as positive change. But the government was less generous than Grant had hoped. Rather than spending its own money, the government used regulations to make others invest in the environment. Grant built his business by tracking the way the money was spent. He watched the industry leap forward on government regulations. And, for the past few years, he has watched the boom stagnate.

"The boom is over," Grant says. "Legacy issues have been dealt with. The 1990 recession started the wake-up-call era. Major new programs ended. Most of the tangible environmental problems were being addressed. What had been front-page news was no longer on the front page. In the 1970s, public opinion polls rating the big issues put environment in the top five, but now environment is in the top fifteen or twenty."

He pauses. "Some people say that stagnation equals success," he says. "We had real problems, and now that those problems are solved we can move on to something else. But environmental stewardship is more than just cleaning up and moving on. Continuing improvement requires monetization of economic incentives. Economics should be considered above all else. It should all be monetized."

Grant returns again to one of his favorite themes. "What is needed now," he says, "is economic consequence." He describes a utopian world where a pollution tax will replace an income tax. In this utopia, taxes are based not on what people produce but on what they waste. But we do not live in utopia.

"The market is not engaged," he says, "and investment dollars will not shift unless there is a consistent economic consequence to unsustainable behavior."

Despite this statement, his own journal shows that the environmental industry—and, by extension, the world as a whole—already benefits from some level of economic consequence. From the pages of the *Environmental Business Journal*: "Accompanying enforcement as a key business driver is the threat of litigation." Another example: "Companies have seen what can happen when irresponsible behavior becomes front-page news. It's no longer just bad press; it's bad investor relations." A third example: "People are finding that, when you do good things for the environment, guess what, you also make money."

"Two to 5 percent is my gut feeling for the number of people informed about environmental economics," Grant tells me. I think about this for a moment, realizing that a figure as high as even 2 percent could only come from someone who spends most of his time with well-informed environmental professionals. I challenge the figure and he agrees. "That may be high," he says, "even for people employed by the industry."

While helpful, while moving in the right direction, Grant believes more is needed. "People need to understand environmental economics. What is needed is a long-term education process. We need to internalize costs." When a power plant emits sulfur dioxide, he believes the plant owner should be billed or taxed based on the amount and constitution of the emissions. The cost of dealing with the damage caused by sulfur dioxide should be billed back to the plant, not externalized to the people who live downwind. Similarly, water pollution and unrecyclable solid waste should be taxed, creating what Grant calls "an incremental negative economic consequence to each increment of unsustainable behavior." Use of land and extraction of virgin material could have similar surcharges. When developers or miners despoil virgin land they should pay a fee based on what he calls "an assessed life-cycle economic value of the land impacted." When someone builds a golf course on top of a wetland, he thinks the owners, and by extension the golfers, should pay for diminished downstream water quality. He sees, too, a need for removal of what have been called perverse incentives—financial rewards for activities that hurt the world.

He thinks these things will happen only if people understand basic environmental economics. If people understand environmental economics, green seduction becomes more seductive. "When we think about how rich we are," he says, "we need to think of our environmental wealth as well as our financial wealth."

This statement circles back to Grant's personal history. As a younger man, he wanted to build solar panels, but Ronald Reagan shut him down. Solar, the Reagan administration believed, would never compete successfully with coal and oil. But the pollution costs of burning coal and oil for power were externalized. Electricity from coal and oil, if their full costs were tracked and internalized, became a fool's bargain.

"Even back in the eighties," Grant says, "if you counted the externalities, wind and solar would have been cost competitive."

This idea—this concept of internalizing costs—is not new. In 1993, Bill Clinton called for measurements of Gross National Product that, in his own words, "would incorporate changes in the natural environment into the calculations of national income and wealth." The year before, the government's Bureau of Economic Analysis, part of the staid Department of Commerce, began looking at what they called Integrated Environmental and Economic Satellite Accounts as a means of incorporating environmental wealth in the tracking of financial performance. In a government-sponsored report by the National Academy of Sciences, an analogy is used to emphasize the importance of tracking the value of environmental goods and services: "Commercial laundry services are reckoned as part of Gross

Domestic Product, while parents' laundry services are not; the value of downhill skiing at a ski resort is captured by Gross Domestic Product, while the value of cross-country skiing in a national park is not." In the United States, Integrated Environmental and Economic Satellite Accounts did not catch on. The power brokers, for now, have not been seduced. But in the Netherlands, in the Philippines, in Japan, systems are in place that attempt to monetize, officially, the environment.

Grant, on certain themes, sometimes repeats himself. "To improve efficiencies," he says, "we need to internalize costs."

This sounds right. Be seduced. Embrace the seduction of green. Sitting here, in the office of the *Environmental Business Journal*, this sounds simple. But where are we now? What keeps our rivers from burning? Why can we still breathe the air? What are we doing, aside from sitting and talking, to move forward toward internalization of environmental costs?

CHAPTER TWO

BUILDING NATURE, INC.

The inside of Mike Rolband's Ford truck smells like wet dogs, and golden retriever sheddings cover the seats and stick to my jacket. Often, the dogs attend site visits with Mike, apparently riding in the front seat, where I sit now, but today the dogs are looking after the office. Mike and I, dogless, drive through a Washington suburb talking about Wetland Studies and Solutions, Inc., Mike's company. We are on our way to Sunrise Valley Nature Park, a wetland park with a boardwalk, made to compensate for wetlands destroyed by the townhouses carpeting this neighborhood.

Mike does not know Grant Ferrier and does not read the *Environmental Business Journal.* They live and work on opposite sides of the country, Grant in San Diego and Mike in Washington, D.C. Grant has a cat, Mike has dogs. In the course of the next day, Mike will say things showing that he is neither politically nor ideologically aligned with Grant. But they share this in common: They have both been seduced.

Mike, in a blue fleece and jeans, eyes hidden by dark sunglasses, leads the way into Sunrise Valley Nature Park. I follow him onto a boardwalk. It is made, Mike tells me, from a recycled product, a mix of discarded plastic bottles and sawdust, with the look of wood but greater durability, especially here, in this wet environment. Chirping birds compete with traffic noise. A jet on its way into Dulles passes overhead. A Canada goose struts across a grassy patch.

Mike says, "The whole site is 15 acres; 3.3 acres is new wetlands, and 1.4 acres is a restored farm pond. Another 5 acres is restored and preserved wetlands. The rest is upland buffers." All of this is code, jargon in the world of wetland regulatory affairs. By new wetlands, he means wetlands that were excavated into existence, a new patch of low ground holding water where previously there was dry ground. A farm pond is a place where a farmer captured a creek behind a simple earthen dam. Restored wetlands are areas that once held water, but that had been drained or filled, probably by the same farmer who built the dam, and that had now been rewetted—undrained or unfilled, so to speak. Preserved wetlands are just that—natural wetlands preserved in their natural state. Upland buffers are relatively natural areas separating the surrounding housing and office buildings from the wetlands.

Mike points out a patch of ground with trees and shrubs. "That's an old farm dam," he says. "It wasn't up to code. The tree roots can compromise the dam's integrity, but we thought the trees looked nice. So we just built another dam behind it." What Mike leaves out is the fact that all of this was done with the eye of an artist. The farm dam does not look like a farm dam—it looks like a patch of trees and shrubs. The new dam blends into the landscape, essentially invisible. I see plants. I see birds. What I do not see is an engineered environment. A person not suffering the effects of too much knowledge could visit this place and feel grateful that a little piece of nature was left behind within the confines of the greater Reston development. This same person might even wonder why the developer showed such constraint, land prices being what they are. And then this person would come across a sign, its words surrounded by colorful drawings of turtles, frogs, tadpoles, and birds: "This wetland area was constructed by Reston Land Corporation to mitigate for impacts to wetlands that occurred during the development of Reston Community."

I stoop to look at a plant. "Soft rush," Mike says, "*Juncus effusus.*" And we move on. He tells me that his employees sometimes come here on their own time. Many of them, like him, have dogs, and they bring their dogs. Mike works seventy to eighty hours a week, but that includes time looking at his projects, almost always with the dogs in tow.

Mike grew up in northern Vermont, cross-country skiing. In the one high school shared by ten surrounding towns, a guidance counselor told him that—because he liked shop class—he should be an engineer. By the time he left Cornell, he had a graduate degree in engineering and an MBA. In 1991 he stumbled into his first wetland project. This was at a time when the nation was struggling to understand Section 404 of the Clean Water Act, the federal legislation that gives wetlands a special status. The trouble with Section 404 in 1991 was that so few people knew how to identify a wetland. Areas that appeared dry could in fact be wetlands. The government developed a manual that allowed specialists and regulators to delineate wetland boundaries based mostly on plants and soils. People were paid to walk around tying ribbons to bushes—on one side of the ribbon, it was a wetland, but on the other side it was not. But even areas that were clearly wetlands did not enjoy anything as simple as total protection. Building in wetlands, or, more accurately, filling of wetlands to build roads, houses, offices, factories, warehouses, or any other artifact of the modern era, required a permit from the federal government. Under the permitting system even a salt marsh could be filled, but only if the government was convinced that no economically feasible alternative existed, and only if the design of the project minimized the amount of marsh that would be affected, and only on the condition that a new wetland would be created or a degraded wetland would be restored to replace the marsh being filled. Developers were up to their hips in the muck of permitting and compliance, mired in wetland regulatory requirements.

Northern Virginia, the suburbs for the nation's capital, was growing, but just as in colonial times the area was wet and muddy, ideal perhaps for politics but less ideal for development, especially under Section 404. In 1991, Mike started Wetland Studies and Solutions. The company's purpose was to help developers through the morass of Section 404. For two years, he ran the company from the basement of his townhouse. Section 404 became increasingly complex, and new wetland laws were passed by state and local governments. As wetland protection grew, Wetland Studies and Solutions grew. Rapidly, it grew to a point at which the only personal space left in Mike's townhouse was his bedroom. In late 1993, Mike moved into commercial office space. Today, Wetland Studies and Solutions fills fifteen thousand square feet of offices and storage. It employs forty-two engineers, biologists, surveyors, and technicians. It grosses over $5 million a year. Bigger companies have tried to buy it. Since startup, it has provided consulting services on more than seventy thousand acres of proposed land developments at more than a thousand sites, and it has created or restored over seven hundred acres of wetlands and restored something like four miles of streams.

Mike's company maintains a map of its sites. The map covers Loudoun, Fauquier, Prince William, and Fairfax counties. Symbols identify sites: yellow duck silhouettes for wetland restoration projects and red fish silhouettes for stream restoration projects. A yellow duck sits on top of Sunrise Valley Nature Park. Scattered across the four-county area are enough yellow ducks to form a flock. Likewise, a respectable school of fish could coalesce from the four-county area. Next to ducks and fish are place names like Bethlehem Assemblages, Dulles Parkway Center, Ashton Avenue, and the Colonies at Scotts Run, functional names associating sites with the construction projects that led to their existence, or, just as often, names that simply state a location. Names ending in "Nature Park" are a minority of exactly one. Next to the county maps, in a small circle, Virginia and West Virginia are outlined. The four counties providing most of the support for Mike's work are shaded, but other than that the map is devoid of detail aside from five ducks and two fish representing scattered jobs outside of his company's normal operating range.

"I have about ten competitors operating in northern Virginia," Mike says, "plus dabblers." Although he is not sure how many wetland companies operate nationwide, one estimate puts the figure at about fifteen hundred—enough to support a flock of ducks that would darken the sky. The quacking overhead would be deafening. Official figures lay claim to about sixty thousand acres of wetland projects each year, managed under ninety thousand permits, to compensate for impacts to something like twenty-five thousand acres. The government claims that the average standard permit application requires just over six months to process, but this is six months after delivery of a completed application package. In the world of real estate development, where timing is everything, there is ample demand for companies like Wetland Studies and Solutions.

"My clients want to obey the rules," Mike says, "but they don't necessarily think that wetlands like these have much value. They just want to build houses. They come to me and say, 'Tell me the laws, and I'll follow them.'" Mike himself has helped write some of these laws, providing suggestions for language so that laws intended to allow responsible development do not simply stop development altogether. He does this, by and large, on his own time.

"One of the problems," Mike says, "is that the system penalizes the honest developer. The guy who plays by the rules hires me, and we submit permit applications—federal, state, and local. The rules don't always agree and sometimes the rules change halfway through a project. Three or four different people, sometimes more, are looking over your shoulder. They're not worried about budgets or schedules. It can be frustrating."

Maneuvering through Highway 50 traffic, headed back to Mike's office, we pass within spitting distance of a dozen sites built by Wetland Studies and Solutions. I know this because of the map with the duck and fish silhouettes, not because Mike mentions them. We pass a ditch overgrown by cattail, and I see a red-winged blackbird clinging to a stalk, its body in the half-twist tilted posture common to these birds when they perch on vertical stems. It occurs to me that this bird has almost certainly spent time in one or more of Mike's projects and that it could be seen, with only a hint of irony, as one of Mike's customers.

By Mike's own account, some people think he is a sellout. They call him scum. They imply that he is only in this for the money, that he helps developers destroy precious wetlands. In at least one case, he was asked to leave a get-together of self-proclaimed environmentalists when he pointed out certain inconsistencies between their lifestyles, which bordered on the opulent, and their stated beliefs, which called for something close to asceticism.

Graduate students sometimes come to Mike in search of employment. "Some of them," he says, "tell me that they will work on restoration projects but not on permits."

At a job site, Mike once scheduled a meeting with leading officials from two government agencies and a rank-and-file regulator who was managing the site's permit. A disagreement had arisen between Mike and the regulator over certain points in a permit application, and the officials were there to settle the argument. Mike stood chatting with the officials while they waited for the regulator. The regulator showed up and marched purposefully across the site, reached Mike and the officials, and verbally attacked one of the officials, calling him "lower than low" and poking him in the chest. Immediately, it became apparent that the regulator had mistaken the official for Mike.

The regulator was reprimanded. Mike got his permit.

From an official at the Environmental Protection Agency's office of wetlands: "Mike is widely regarded among the federal agencies as someone who makes the extra effort and spends the extra money to do it right. I don't know of anyone who thinks of him as a sellout."

A land-use attorney, sporting hiking boots and a trench coat over his dress shirt and tie as he and Mike hiked around a parcel of land in northern Virginia, once said of Mike, "He can wear fatigues all he wants, but he's a capitalist. God bless him."

Mike has awards from the Engineers and Surveyors Institute, the Home Builders Association of Virginia, and the National Association of Industrial and Office Properties, but he also has awards from Maryland's governor for Chesapeake Bay Watershed Improvement and from the Boy Scouts.

"I guess I'm competitive," Mike tells me. And, in an unrelated conversation, he says of his company, "We just want to be the best at what we do. Period."

Wetland Studies and Solutions works out of rented space in a Chantilly, Virginia, office park. Above the office, pinned to the brick wall above a second floor window, the word "Wetland" stands out, three times bigger than the rest of the company's name, Studies and Solutions, Inc. Mike's two golden retrievers, Woodrow and Truman, serving as receptionists, show me in. I talk briefly to a woman who is headed into the field to mark a wetland. She has a PhD, and she is a Registered Professional Forester and a certified Professional Wetland Scientist. She has additional training in soil taxonomy, storm-water management, computer-assisted drafting, and geographic information systems. In any one week, she will draw from all of this training, as well as her fifteen years of experience in the environmental arena. She is not atypical of employees at Wetland Studies and Solutions, Inc.

I follow Woodrow and Truman to Mike's office, where he stands bent over a drafting board with one of his staff. They are discussing the plan for a low-density housing project in Fairfax County, thirty minutes southeast of here. The plan sports two sets of lines delineating Resource Protection Areas, or what Mike calls "RPAs," where building will be restricted to protect stream corridors and wetlands. One set of lines shows restricted areas as they stand today, another set shows restricted areas as they are likely to stand after new laws take effect. Under the new laws, the developer will lose about half of his currently usable property.

Mike's degrees hang on the walls of his office, but so do pictures of his dogs. A white board covers one whole wall, with a complicated chart of employee's names linked to dates and tasks. A bin in one corner is filled beyond capacity with rolled maps. Two chew toys lie on the floor. The overall effect is one of controlled clutter. Mike offers me a chair, and before I can sit both Woodrow and Truman take up stations on the floor, next to their chew toys, both eyeing me, tongues out, panting.

Sitting, Mike explains some of the concerns surrounding projects like the Fairfax low-density housing development. He explains that typical small projects—including rezoning, site planning, and construction—take three to four years to approve and build. Bigger projects require more time. Regulations can change once or twice a year, and agency interpretations change more frequently.

"We started working on the Reston project in 1993, and it's approaching completion. But the project itself started in 1963. It will be over forty years to complete that community. The Broadlands is another example. We started that project in 1994. It's maybe five thousand houses, and after ten years it is maybe one-half constructed. The average life span of the big, planned community projects that we work on is maybe fifteen to thirty years."

Among Mike's clients, the uncertainty associated with wetland rules, as much or more than the rules themselves, leads to what Mike calls, with a smile, "extreme frustration and anger." He tries to get permits for the longest possible period—fifteen years under current laws for state and federal permits. There are also local regulations. "Loudoun County, for the last two years one of the fastest growing areas in the country, passed some very, very tough regulations. On a typical piece of land in Loudoun County, our billing for wetland and natural resource work quadrupled."

Leaving aside state and local regulations, federal regulations alone could bog down the most intrepid developer. One might ask how the Corps of Engineers, run by the military, became responsible for wetland permitting. The answer is not straightforward. It came about in the 1970s. The Corps, responsible for maintaining navigable waterways, had been given the authority to regulate placement of waste materials—primarily dredged sand and mud—since before the start of the twentieth century, under Section 13 of the Refuse Act of 1899. This authority was limited to navigable waterways. Richard Nixon directed the Corps to convert this authority into a program that would "regulate the discharge of pollutants and other refuse matter into the navigable waters of the United States or their tributaries." Certain members of Congress—notably Edmund Muskie, a Maine Democrat—did not want the Corps in the business of environmental regulation. Ultimately, a compromise was worked out whereby placement of dredged sand and mud would be regulated under Section 404 of the Clean Water Act rather than the Refuse Act of 1899, and the Environmental Protection Agency would develop guidelines for the Corps. The compromise also changed "navigable waters" to "waters of the United States." In a suit brought by the Natural Resources Defense Council, an environmental advocacy nonprofit organization, the Corps was forced to include wetlands in their definition of "waters of the United States," whether or not those wetlands could even float a canoe. Since then, there have been hundreds of court cases, usually pitting environmental advocacy groups against developers, in a give-and-take relationship that has looked at everything from the types of wetlands that can be protected to whether or not permits are needed to regulate dirt that accidently falls into a wetland off the edge of a front-end loader. Often, the regulatory process leads to requests for public comment on specific issues. Recently, in response to a request for comments regarding the regulation of small wetlands that are isolated on the landscape, the Environmental Protection Agency received more than 135,000 comments, generating more than enough paper to fill a small wetland. People are paying attention.

"Since 1991," Mike says, "things have become much more complicated and way more difficult. In 1991, you could impact ten acres under Nationwide Permit 26 without a lot of problems." There are forty-nine so-called Nationwide Permits, or Nationwides, which are rules applied throughout the country to cover commonly occurring specific activities that are perceived to have only minor impacts. Routine activities like road maintenance can sometimes be undertaken using one of the Nationwide permits. When they work, the Nationwides streamline the regulatory process. "Under one acre," Mike says, "you could impact without mitigation. Nationwide 26 is no longer used. Now you need mitigation for anything bigger than a tenth of an acre. And in some areas, some of the Nationwides have been suspended and replaced by state programs. Things that were permitable then you couldn't do now. And easy things then, things that we would have charged a few hundred dollars for, are now several thousand dollars."

"It seems to me," I tell Mike, "that the laws get tougher as population densities increase."

"At the local level, that's true. But at the national level, I'm not so sure. Pressure groups affect the process. A lot of these regulations come from the Bush administrations. It was the first Bush administration that pushed for no net loss of wetlands through restoration, and this last crackdown came under the second President Bush. Most of our clients still can't believe it. The majority of clients are conservative Republicans, but they get hurt by Republicans, even at the state level. It was easier to get a permit under Governor Wilder, who was a Democrat. When Governor Allen took over he decimated the ranks of the regulatory program. Now it takes way longer to get a permit than it used to."

Enter into this landscape Wetland Studies and Solutions and hundreds of companies like it, niche companies responding to a demand created by a single aspect of the environmental legislation that exists in the United States. Mike has a notebook that he sometimes gives away, full of descriptions of his company's services, employee resumes, and blurbs about past projects. The outside cover shows a boot print, and next to it the words, "There's no one better in the field." The inside cover, which shows the boots presumably responsible for the footprint, says, "How can a group of scientists, engineers and surveyors in muddy boots help your bottom line?" And for clients the bottom line is what this is all about.

"We started the Broadlands project in 1994," Mike says. "The original plan would have impacted over forty acres of wetlands. Mobil Land was interested in buying the project, so they had a three-day meeting, Friday through Sunday, to think it over. We redesigned the project so that only eleven acres were impacted. We did most of the mitigation on site, and we set up a trail system and a greenway corridor through the site that became a key element of their marketing program. They've had ads in the *Washington Post*."

Another example, but without the happy ending: Mike set up an opportunity for the Virginia Department of Transportation to buy into one of his projects at about eighty thousand dollars per acre. Instead, because of requirements from the Corps of Engineers, the Department of Transportation spent, to date, one million dollars per acre on a restoration project that leaves Mike, as he puts it, "skeptical." A third example: The government, wanting to build a civil works project, exercises eminent domain to forcibly buy someone's property and applies a very low price per acre, arguing that the property cannot be developed because it is covered by wetlands; Mike's company comes in to demonstrate that permits could be obtained. The value of the land increases by virtue of a couple of meetings.

Stories like these bring in clients. In 2002, Wetland Studies and Solutions had 230 separate clients. Of these, about 90 percent are repeat clients. The biggest client accounts for only 5 percent of the company's total revenue, and the next biggest client accounts for less than 3 percent. One hundred and sixty-six of these clients spent less than ten thousand dollars each—small jobs, but lots of them. He attracts and holds clients by being active and visible in building industry trade associations. "We help them when any type of regulatory issues comes up. Right now, my personal focus is in Fairfax County, where the Chesapeake Bay Act is being changed. I'm going to all the workshops and testifying at hearings, working on affecting the legislation. And we use our newsletter to keep clients informed and attracted to our business."

In the newsletter, a headline—all capitals, underlined, and bold-faced—reads, "Regulatory Changes Coming at All Government Levels." The article under the headline warns of "major regulatory changes that are causing significant effects on all building industry and public works projects." These changes will lead to an "order of magnitude cost increase for the mitigation of impacts to streams" and a "two-fold, or more, increase in the area of Resource Protection Areas." Another headline reads, "Virginia Pollutant Discharge Elimination System and Chesapeake Bay Preservation Ordinances—They Say Change Is Good?" followed by an article explaining that an existing regulation "has been amended to reduce the amount of land disturbance requiring a general permit from 5 acres to 1 acre" and that, with this amendment, "any ongoing 'small' construction activity that began before December 4, 2002, is now required to submit a Registration Statement." Mixed with this are occasional passages that could have fallen from the pages of Poor Richard's Almanac: "Don't complain. Be Proactive!" and "The key is participation." With the exception of a few short sentences inserted at the end of each newsletter, there is no advertising touting Wetland Studies and Solutions, Inc. There is nothing about how Mike's projects benefit the environment, or society, or his clients.

We eat lunch while we talk. The smell of sandwiches excites Woodrow and Truman. Mike turns his back for a moment, and Woodrow pulls the sandwich from Mike's desk. Both dogs are immediately evicted. Mike picks his sandwich up from the floor, reassembles it, and eats it.

I ask Mike if his clients are familiar with work by the economist Robert Costanza, director and founder of the Gund Institute for Ecological Economics at the University of Vermont, who claimed that wetlands are worth just under six thousand dollars per acre per year and that investing in the preservation of intact ecosystems yields one-hundred-fold returns on investment. These claims were not made in left-wing environmentalist's journals, but in the prestigious academic journals *Nature* and *Science*, where articles cannot be accepted without surviving the rigor of strict peer review. Neither Mike nor his clients, as far as he knows, have ever heard of Robert Costanza. But this does not mean that Mike does not have a view about what wetlands are worth.

"I look at real estate values," Mike says. "Sally Mae just bought four acres of land for eight million dollars in Reston Town Center. If you're in an area where land is worth forty dollars a square foot, the wetland is worth forty dollars a square foot. If you're in an area where land is ten cents a square foot, that's all the wetland is worth. It's not about ecological function. It's about what people will pay for it."

Mike and his clients may not be among the 2 percent of the people whom Grant Ferrier believes to be informed about environmental economics, but they are among those impacted by internalization of environmental costs. Earlier, Mike had talked about "extreme frustration and anger" among his clients, but this kind of money, in this kind of setting, breeds nothing short of fear and loathing.

"The majority of time," Mike says, "developers and environmental groups view one another as the enemy. It's easy to become cynical about environmental regulations. The same governments that issue permits—the ones that set up the toughest regulatory environments—go out and build a school or road or some other government building on a wetland without a second thought. There's a double standard."

One of the dogs scratches at the office door. The sandwiches are gone, and Mike lets the dogs back in. Someone has put a rubber band around Woodrow's snout, an undignified accessory for a golden retriever. With his soft laugh, Mike slips the band from Woodrow's snout and rubs his head. Woodrow, looking for a moment remarkably like a teenager in love, stares up at Mike.

Mike's office building does not have a green design. It is a standard rental property. The retirement fund for his employees is not a green investment mutual fund but, rather, the standard retirement fund used by a homebuilders association. His company owns a Toyota Prius, the hybrid car that combines gasoline and electric motors with advanced technologies to deliver fifty-two miles-per-gallon fuel economy. Toyota asks customers, "How far will you go to save the planet?" and answers for them, "About 566 miles per tank." But Mike, by his own account, bought the car not so much to save the planet as to save on fuel costs, and, importantly, to take advantage of rules that allow anyone in a hybrid car to use the "high occupancy vehicle" lanes normally reserved for carpoolers, allowing both him and his employees to zip across town to meetings with less downtime for travel. When a hybrid pickup truck or utility vehicle comes onto the market, he intends to buy one of those, too, for the same reason.

We talk about these things while Mike, Woodrow, and Truman show me around the office building. An employee is introduced to me as a regulatory specialist, another as a floodplain guru, another as a surveyor. About half of the offices are empty, their occupants in the field. There is almost no place in this building from which a map of one kind or another is not visible. I poke my head into what Mike calls "the library." Scattered amongst binders of photocopied environmental regulations, there are technical books: *National Environmental Policy Act Law and Litigation, Applied River Morphology, Creating Freshwater Wetlands*. We walk outside and around the corner to another part of the building, a large room with a concrete floor and bare cinder block walls, used to store the tools of the trade. There I see a small tractor, two all-terrain vehicles, a pair of backpacks full of electronics used in surveying, boots, shovels, racks of PVC pipe.

A blue canoe and a green johnboat hang from the ceiling. Large white bags of seed form a pile three feet high and twenty feet long. Nearby, a metal rack holds reports generated by Wetland Studies and Solutions, filed in white and blue cardboard packing cartons and neatly labeled. There are more than three hundred cartons, stacked four high.

Soon after he started Wetland Studies and Solutions, Mike saw an opportunity with a company called South Charles Realty, what he refers to as an REO, a real-estate-owned division of a bank. REOs deal with properties acquired through foreclosure. The company contracted Wetland Studies and Solutions to assess wetlands on its properties. "We also did a lot of garbage pickup for them," Mike says. "Things were tough back then."

In Prince William County, one of the REO's sites consisted of 227 acres adjacent to Neabsco Creek. The bank had loaned something like four and a half million dollars on this property with the belief that more than two hundred houses could be built. Mike showed them that 201 of their acres were wetlands. Almost all of what was left was in the floodplain. The original loan was made when wetlands could be filled in, and the owner had planned a large subdivision of single-family homes. But things had changed since they made the loan. The property was, for all practical purposes, unbuildable. Too much of it was covered by wetlands.

"One of the ideas I had," Mike tells me, "was to make more wetlands there. It was the perfect combination of real estate development and doing better things for the environment. I read as much as I could about wetland banking, which was at that time a new concept." Wetland banking works like this: Wetlands are restored, created, or preserved on a piece of property, which then becomes a bank with wetland credits that can be sold to other developers who are destroying wetlands in the immediate vicinity of the wetland bank. The clients for the bank credits are developers looking for certainty, looking for assurance that they can hit their wetland permit requirements.

"The Corps representative loved the idea," Mike says, "but we had huge problems with the Environmental Protection Agency." Certain individuals at the agency did not like real estate speculators turning to wetlands as a profit center. A banking agreement was signed by the Corps, without the Environmental Protection Agency's concurrence. This was a big risk on Mike's part, in that the Environmental Protection Agency could try to block sales of credits from the wetland even though the Corps had approved. Mike formed a partnership with the bank—the real bank, the financial institution that owned the REO and its land—to build the first phase of the wetland bank. He presold just over two acres, "a chunk of it," to a company building affordable housing for the elderly. He had to work with the state to develop legislation establishing what he calls "the service area"—the area of land, surrounding the bank, in which he could sell wetland credits to other developers. The first phase was approved at the end of 1994, but the second phase was not approved until after the state law changed, in 1996, five years after the project's inception.

The Corps was so impressed that they asked Mike to name the site after Julie J. Metz, a government biologist who played a key role in developing standards for wetland banking before dying of breast cancer at thirty-eight years old. Her ashes were spread nearby. Ultimately, the Julie J. Metz Wetland Bank provided wetland credits to sixteen separate real estate development projects, after which it was donated to Prince William County as a park. Birdwatchers have recorded more than 160 species of birds on the site. There are more than two miles of trails and a blind for observing wildlife. The site has become part of the Virginia Birding and Wildlife Trail. Visitors from Vietnam, Japan, the Terrene Institute, and Congress have toured the site, recognizing it in part for its natural beauty, but really because it links entrepreneurial practices with nature conservation.

Since then, Mike has built three more wetland banks. One provided mitigation for more than forty projects before it was donated to the National Capital Area Council of the Boy Scouts of America to become part of Camp Snyder. Another, now less than three years old, hosts thousands of frogs, more than 100 species of birds, and an increasingly diverse plant community, all on what had been cattle pasture and crop land.

It no longer matters, but the Environmental Protection Agency never signed off on the Julie J. Metz Wetland Mitigation Bank. Outside, fear and loathing continue to permeate the world of wetland permitting and mitigation. But here, in Mike's office, Woodrow and Truman lie panting on the floor, watching attentively as Mike finishes this story.

Mike, accompanied by a lawyer and a property owner, walks across a vacant lot next to an office park. It is close enough to the nation's capital to be worth more money than most people will earn from a lifetime of labor. The property, recently acquired by Mike's client, will eventually support an office building. At a minimum, several hundred employees will work here. But there is a catch: A stream runs across the property. The stream's babbling mixes with traffic noise. At its widest, the stream can be crossed by a long stride. Because of the property's configuration, it will be impossible to erect an office building without impacting the stream. With this in mind, government agencies allowed the previous owner to fill the stream with rocks. Beneath the rocks, a plastic pipe conveys most of the stream's flow. The stream flows freely onto the property, enters the pipe buried beneath the rock, flows through the pipe, then, at the other end of the property, leaves the pipe to flow freely. The pipe is perforated, so that runoff from the property can flow through the rock and into the pipe. The stream reminds me of certain streams that disappear underground in Florida and the Yucatan, then reappear a few hundred feet downstream, or a mile downstream, or seven miles downstream. The difference is that this one involves plastic and imported rock. The approach is very similar to that of Mike's Stonecroft Boulevard site, where storm water from the parking lot was piped away, underneath the site, to the wetland that Mike had built.

Here is the catch: The permit that allowed this stream to be piped and filled has just expired. And, to make matters worse, it appears as though the previous owner may not have installed enough rock to fully satisfy the permit conditions. Rock should have been installed to the top of the stream banks, so that all of the water's flow, during a strong and steady storm, would be conveyed within the pipe and the rock, without overflowing. Instead, just enough rock was used to cover the pipe. Even though all of this was done before the property changed hands, the regulatory agencies are under no obligation to extend or renew the permit. They could refuse to allow more rock to be placed in the stream. They could stop the project altogether.

Very quickly, through an exchange of ideas between Mike, the lawyer, and the property owner, a solution emerges. The lawyer will approach the agencies with an offer to support a tree-planting and nature-trail construction program in a nearby park in exchange for permission to complete this project. The cost of the program will be small relative to the value of the property, but large in view of the fact that this project had already been permitted once. Both Mike and the lawyer seem reasonably certain that this offer will be accepted. The property owner appears relieved. The employees who will one day work on this site know nothing of this plan, but if they did, they, too, would be relieved.

Not too far from this site, Boy Scouts paddle a canoe in a wetland that would not exist but for Mike. The Boy Scouts may, at times, be lucky enough to see Woodrow and Truman splashing through the wetland, chasing sticks thrown by a bearded man walking along the bank. The boys will develop childhood memories of Mike's wetland. Some will become nature lovers. Almost certainly, some will grow up to object to the very idea of permitting the destruction of natural wetlands in exchange for new wetlands made by human hands. But others could eventually find themselves working in the building that will one day be erected on the site with the buried stream, a building that would not exist but for Mike.

CHAPTER THREE

STOP AND THINK

Things do not always go as one might hope. The Crown Jewel Mine, 160 miles northwest of Seattle, was undeniably controversial. Knowing that it was controversial, the Battle Mountain Gold Company, which owned the Crown Jewel Mine, had retained expertise to gather and interpret the biological facts surrounding the project. And knowing that it was controversial, they were not unduly surprised by the billboards along the road decrying the mine. Likewise, they were not surprised by the standing-room-only crowd at a public meeting about the mine. On the other hand, the people in costumes—six-foot-tall owls and mountains with legs—were a surprise. The chanting was a surprise. The singing was a surprise. The crowd's refusal to engage in reasoned discourse, while perhaps not a surprise, was an annoyance. Watching these events unfold was the equivalent of watching a train derail: The proposed mine's environmental impact statement, required by the National Environmental Policy Act, may have just jumped the tracks.

Or maybe not. "This is typical behavior at some of these meetings," Stuart Paulus tells me, "and it really has only limited bearing on the outcome of the environmental impact statement." Stuart would know. Although he had once scratched out a living studying alligators and birds and publishing scholarly papers with titles like "Time Allocation by Aleutian Canada Geese during the Nonbreeding Season in California" and "Feeding Ecology of Gadwalls in Louisiana in Winter," he now produces environmental impact statements.

But whether this was typical behavior or not, whether it had limited bearing or not, the Battle Mountain Gold Company ultimately abandoned plans to open the Crown Jewel Mine, leaving in the ground more than ninety thousand pounds of gold. The residents of the surrounding area would not enjoy the benefits of more than a hundred new jobs, and both the Bureau of Land Management and the U.S. Forest Service would forego revenue from mining. No access roads would be built; there would be no processing buildings, no assay buildings, and no maintenance shop. More than a square mile—787 acres of land—would be spared. There would be no risk from the cyanide that would have been used to separate the gold from the ore. There would be no impacts to streams and no change in scenic values. Wildlife would be spared the grinding noise of the mill and the steady hum of machinery.

Stuart's role in all this had been that of a subcontractor, working for a large consulting firm contracted by the Battle Mountain Gold Company. "Issues came up related to wildlife," he says. "What are the impacts to deer? So I did a deer study. What are the impacts to amphibians? So I did an amphibian study. What are the impacts to bats? So I did a bat study." He served as an expert reviewer and as an expert witness. In his view, the impacts to wildlife were less than devastating. Other suitable wildlife habitat was plentiful in the surrounding area. Some species would be attracted to the mine, or at least its fringes, just as they are attracted to natural breaks in the forest. And the scar from the mine, when it closed, would slowly fade.

What happened? During the development of the environmental impact statement, a multivolume report—prepared, according to the authors, "to inform the federal, state, and local agency decision makers of the probable environmental impacts"—attracted not only attention, but fear and loathing. The mine had many supporters, but the detractors were more vocal and more organized. By virtue of their enthusiastic din, detractors appeared more numerous than supporters. By the time the environmental impact statement had been completed, few people really cared what it said. You were either for the mine or against it, facts of the case notwithstanding. The mine—a one-square-mile piece of land in a state with seventy-one thousand square miles—became politicized. Before it was over, certain proponents of the mine chastised Stuart, with tongue only begrudgingly in cheek, for not pushing certain environmental activists over the edge of a certain cliff. A senator ended an eighteen-year run in Congress, in part because he supported the Crown Jewel Mine; he had, among other things, attached wording that favored the Crown Jewel Mine to a bill whose main purpose was to fund Kosovo refugees and victims of Hurricane Mitch. The Battle Mountain Gold Company reputedly spent close to eighty million dollars on a project that would not produce a single ounce of gold.

"My strength," Stuart tells me, "is in negotiating win-win situations for conservation and economics." On another project, this one called the Pend Oreille Mine, Stuart's strengths hit pay dirt. The Pend Oreille Mine was smaller than the Crown Jewel

Mine: It would scar only about a hundred acres of land. It was a tunnel mine rather than a surface mine. It was further east, near the border with Idaho, in a district known for mining and logging. Rather than gold, the Pend Oreille Mine would remove six million tons of lead and zinc ore over an anticipated eight-year life span. Rather than building a new mine, the project would renovate and expand existing mine shafts. It would use existing processing buildings, shops, and access roads. On the other hand, it sat on the banks of the Pend Oreille River, where it could potentially affect long reaches down stream.

In this case, Stuart, through his current employer, contracted his services directly to the government. This employer is ENSR, one of the world's largest environmental consulting firms. For companies like ENSR, large government contracts are a mainstay. So, rather than just conducting wildlife studies, his task was to manage and help write the entire environmental impact statement.

"Mining is centered in the eastern part of the state," he says. "Mining and timber are big job generators over there. People around Pend Oreille are a little more sensitive to the resource extraction industry. The government manager in charge of the Pend Oreille environmental impact statement had a basic philosophy—he wanted the project to live or die on its merits and not on peoples' perceptions about mining."

In his Seattle office, Stuart compares the Pend Oreille and Crown Jewel projects. "A public relations firm was hired to unroll the Crown Jewel project," he says. "Some of the things they did for public relations would have ticked anyone off, and their adversaries spun it right around. Here's an example. The public relations firm released a brochure that started with, 'Some people say this is a red herring.' Their adversaries turned it around, saying, 'Not only will Crown Jewel kill the red herrings, every other fish will die, too.' Then the red herring became a symbol for some of the anti-mining groups. They used it as a logo." Stuart stops, reflecting for a second. "On the Pend Oreille project, we engaged the public and known mining adversaries right from the beginning."

The mining company asked some of the anti-mining groups if they were willing to work together to make Pend Oreille a good project. Some of the same people who killed the Crown Jewel Mine agreed to work with Pend Oreille and, possibly, to even help out with some of the monitoring. "The company," Stuart says, "wanted them to feel welcome on the site. They wanted the anti-mining groups to make sure things were okay."

During the Pend Oreille public meetings, things went, more or less, as well as one might hope. No one showed up dressed as owls or mountains. No one sang or chanted. In 2003, Pend Oreille Mine began operations. Stuart's management led to the first metals mining environmental impact statement in ten years that avoided a significant legal challenge in Washington state. Whatever environmental impacts will come from mining at Pend Oreille are the accepted tradeoff, allowing lead for batteries and zinc for galvanized nails.

A Web search with the key words "environmental impact statement" will return more than seventy thousand hits. There is the "Final Environmental Impact Statement for a Geologic Repository for the Disposal of Spent Nuclear Fuel and High-Level Radioactive Waste at Yucca Mountain, Nye County, Nevada." There is the "Adult Mosquito Control Programs Draft Environmental Impact Statement" of the New York Department of Health. There is the "Final Environmental Impact Statement for the Interagency Bison Management Plan for the State of Montana and Yellowstone National Park." Nationwide, an average of about two environmental impact statements have been completed every day since 1979. The National Environmental Policy Act, or NEPA, signed by Nixon on New Year's Day, 1970, requires environmental impact statements for all federal government activities that could have substantial effects on the environment. This includes the obvious, like new interstate highways and the construction of navigation channels, but it also includes any action requiring a federal government permit, any action requiring access to government land, and any action changing the way a federal entity performs its duties. If the Bureau of Land Management changes weed control methods in a way that might have significant effects, an environmental impact statement is needed. If the army changes training practices, with possibly significant implications, an environmental impact statement is needed. If the Forest Service changes timber management protocols in ways that might hurt the environment, an environmental impact statement is needed.

But Nixon was no tree hugger. When he signed off on NEPA, he may have been somewhat distracted. He had recently ordered secret bombing raids into Cambodia while simultaneously considering troop withdrawals from Vietnam, the My Lai massacre had just hit the press, the Chicago Seven were turning their trial into a circus, he was thinking about the possibility of a nuclear arms summit with Leonid Brezhnev, and he was worried about China. NEPA probably appealed to Nixon. It appeased the environmentalists, who at that time were part of a relatively small but vocal and obviously growing movement, and it created the Council on Environmental Quality, which would advise the White House on environmental issues. He needed advice. Environmental problems were real. That very year, Ohio's chemical-laden Cuyahoga River had once again burst into flames. Four years earlier, Clair Patterson testified to the Senate about apparently deliberate falsehoods in lead industry research that justified ongoing use of lead in paint and in gasoline. Seven years earlier, Rachel Carson's *Silent Spring* had ushered in a groundswell of well-justified public concern about pesticide use. In the midst of these realities, NEPA's wording, to someone like Nixon, in a position like Nixon's, must have appeared benign: "It is the continuing policy of the Federal Government, in cooperation with State and local governments, and other concerned public and private organizations, to use all practicable means and measures . . . to create and maintain conditions under which man and nature can exist in productive harmony, and fulfill the social, economic, and other requirements of present and future generations." It addressed the issues of environmentalists, but framed itself around the economic needs of people, and with little pain it freed Nixon to get back to what he almost certainly thought of as more serious business. And it is easy to imagine that he liked the wording, that he liked, in particular, the phrase "productive harmony."

Within the environmental arena, NEPA is an odd law. Most environmental laws either restrict or encourage specific behaviors. The Endangered Species Act restricts actions that harm endangered species, while the Resource Conservation and Recovery Act requires specific protocols for the generation, transportation, treatment, storage, and disposal of hazardous waste. In contrast, NEPA is so general as to be little more than a statement of philosophy, at least at first glance. It talks of protecting the environment for "present and future generations of Americans," of assuring "aesthetically and culturally pleasing surroundings," of supporting "diversity" and "variety of individual choice," and of the importance of "a broad sharing of life's amenities."

As the environmental movement evolved, NEPA's number one tenet became the requirement for an environmental impact statement. It works like this. The federal government—meaning a person or a group of people working for the federal government, most of them sitting in cubicles and wearing plastic identification badges, some of them disgruntled about low salaries and stifled careers—makes a decision. The decision could be to lease a block of land for logging, to change a policy about nuclear waste storage, to dredge a navigation channel, to restore a wetland, or to build a new stretch of highway. For projects that might have environmental effects, an environmental assessment is undertaken—in some ways, a pared down environmental impact statement. If it is determined that the project is relatively benign, a "Finding of No Significant Impacts"—also known as a FONSI—is filed, and the project moves on. If significant impacts appear possible, a full-blown environmental impact statement is required. A report, often running hundreds of pages and almost always unreadable, is generated. The report discusses alternative versions of the proposed project. While the report relies mostly on existing information, additional information needs are sometimes identified and new research can be required. Public notices must be issued and public meetings held. Comments are solicited from all interested parties and these comments must be considered. The process can drag on. All of this is done under the auspices of a government agency, but the work itself and even the management of the work is often contracted out to private-sector consultants like Stuart Paulus. And, when the decision involves a private-sector entity—a mining company applying for federal permits or hoping to use federal land, for example—the private-sector entity pays the bills. When the dust settles, the end result is a Record of Decision—in jargon, a ROD.

But here is the catch: The Record of Decision carries no weight. NEPA was designed and intended to force consideration of environmental effects, but it was not intended in and of itself to see that anything was done about these effects. People call NEPA the "Stop and Think Act." While it requires that we think about our actions, that we consider the environmental consequences of a plan before that plan is brought to fruition, NEPA does not say that a plan must be abandoned or altered if it will cause environmental devastation. The next step, the critical step, is incorporation of recommendations laid out in the Record of Decision into permits required by other environmental laws or lease requirements allowing use of federal lands.

It has been said, more than once, that NEPA has no teeth. Twenty-five years after the law had been passed the White House itself was concerned enough to publish a report looking at the effectiveness of NEPA. A great deal had happened in those twenty-five years. Nixon, of course, was gone. The United States population, and with it the potential to impact the environment, had picked up ninety-five million people. The environmental movement had become big business, with nonprofit organizations using budgets of tens of millions of dollars to fight environmentally destructive development, with hundreds of environmental consulting firms stepping up to the plate to help industry and government deal with environmental issues, with every major company supporting an environmental permitting and compliance team, with universities churning out students from environmental studies programs that did not even exist when NEPA was signed. The Internet had come to life, allowing information to flow like never before and giving real opportunities for public involvement on a massive scale.

According to the White House report, "NEPA is a success—it has made agencies take a hard look at the potential environmental consequences of their actions, and it has brought the public into the agency decision-making process like no other statute." It describes NEPA as "the foundation of modern American environmental protection." It claims that "interagency coordination under NEPA has avoided or resolved many conflicts, reduced duplication of effort, and improved the environmental permitting process." It reports that "NEPA has been emulated by more than 25 states and 80 countries around the world, and serves as a model for environmental impact assessment for such global institutions as the World Bank." But it goes on to say that "NEPA's implementation has sometimes fallen short of its goals." It points out that some see the environmental impact statement as an end in itself, suggesting that the law sometimes results in nothing more than a report destined for a bookshelf, a process generating a short-lived sound and fury that in the end signifies nothing. It says that "agencies sometimes engage in consultation only after a decision has—for all practical purposes—already been made."

Does NEPA have teeth?

A man named Greg, commenting about a specific project: "The environmental impact statement route was taken precisely to put NEPA teeth into things."

From the National Academy of Public Administration: "The subsequent development of environmental impact statements as a vehicle for public access to agency decision making has given NEPA some 'teeth' while also making it controversial and challenging for agencies to implement."

Walter W. Olsen, in a textbook chapter for engineers: "The teeth in NEPA lays in the subsequent volume of lawsuits that has followed its passage. Federal projects of all manner and nature have been challenged under NEPA, many successfully. Others have been delayed based upon NEPA challenges."

A man named Jeffery: "NEPA is a paper tiger. Why is the Supreme Court pulling the tiger's teeth?"

A man named Steve: "I'm sorry if I was misunderstood when I said that about NEPA, but in the context in which it was stated it still holds true. What I meant by 'no teeth' is that it is not enforceable through fines or stipulated penalties, nor is it a criminally enforceable statute, as unlawful disposal of hazardous waste might be. I meant nothing more and nothing less."

A woman named Marianne: "No need to check with a lawyer. I have successfully used NEPA on many occasions to stop government actions. Therefore, I believe you should rescind your statement . . . that NEPA has no teeth. It just isn't true."

From a government report: "Citizens sometimes feel frustrated that they are being treated as adversaries rather than welcome participants in the NEPA process."

A civilian employee of the federal government: "NEPA is about balancing public outcry against doing as little as possible."

NEPA, questions about the sharpness of its teeth notwithstanding, has accomplished a great deal. Although some might question NEPA's efficiency, no one working in the environmental arena would seriously suggest that it has not contributed to improved environmental stewardship in the United States. And, unquestionably, it has created something that goes way beyond a cottage industry: It has led to hundreds of millions of dollars worth of work for environmental consulting firms. According to Grant Ferrier, the demand for professional environmental consulting services—the services needed to generate an environmental impact statement—has grown by a factor of fifteen in about twenty years.

Late on an overcast Monday morning, the ENSR parking lot is all but empty. Through the drizzling rain, looking only at the sign above the door of this midsized gray office building in a light industrial area outside of Seattle, there would be no way to know exactly what ENSR does. But knowing the industry helps. ENSR, short for Environmental Services, is the world's sixth largest environmental consulting firm. It employs over thirteen hundred scientists, engineers, and project managers working from more than seventy offices on projects in more than forty nations. It has completed more than fifty thousand environmental consulting projects. Annual gross revenues exceed $150 million.

Upstairs, books line one wall of Stuart's office: *Mammals of the Pacific States, Birds of North America, Wildlife of the Northern Rocky Mountains, Freshwater Invertebrates of the United States.* A Crown Jewel Mine ball cap, white with gold embroidery, sits on a shelf. A signed baseball and several golfing plaques stand behind his desk, and the desk itself doubles as a filing cabinet, holding stacks of papers, each representing one aspect or another of one project or another that Stuart happens to be managing.

"Things are quiet this morning," he tells me. "We're sensitive to economic trends. It's the nature of the business. We laid some people off in the last year or two. In this office, we have about twenty-five or thirty people now, down from about sixty when I started. A few years ago Enron was one of our biggest clients. Obviously, we took a major hit from that and people moved around."

Stuart himself is an example. His bachelor's degree came from the University of California at Davis, his masters from the University of North Dakota, and his PhD from Auburn University. After finishing his degrees, he worked for a time at Louisiana's Rockefeller Wildlife Refuge, living in a town that he characterizes as "a water tower and a fire station." He moved west, away from his ducks and alligators, and, he hoped, to greater financial rewards, applying some of the computer-intensive statistical methods he learned as an ecologist to financial market analyses. Five years later, missing fieldwork, he was drawn back to the environmental arena, seduced back to what he refers to as his "roots." He worked for a small firm, like Mike Rolband's company, specializing in wetlands.

The move to the world of environmental consulting was not necessarily an easy transition. "The folks I worked with in Louisiana hated environmental contractors. They thought of environmental contractors as scum of the earth who would help their clients develop land at any cost. So I started off with a bad impression. And some of my work at that time was for small developers. There was pressure to cut corners. They would lay out a plan for a property and then ask me to make it fit the regulations, and we would spend our time arguing. I had to repeatedly explain that the law would not allow what they wanted. Some of them wanted us to bend the rules. I wouldn't do it."

Before long, by chance, the firm he worked for assisted ENSR on a project, as a subcontractor. ENSR liked him and he liked ENSR. He liked the independence he could find with a larger company. He liked working with larger clients who did not apply pressure to cut corners, who did not feel that they would go out of business next week if a permit failed to come through as planned.

"Development is going to happen whether you like it or not," Stuart says, "but it can be done with a careful and well-thought-out approach. It can be a win-win."

Now, with ENSR, he seems at home, offering ecological expertise for a company that provides a full suite of environmental consulting expertise. Emphasizing this, just down the hall from his office, a window looks into another part of the building currently housing a two-story-tall plumber's nightmare, a compilation of pipes, tubes, pumps, and tanks used by engineers to model water flows. It might be said that such a contraption is anomalous next to an ecologist's office, or that the ecologist's office is anomalous next to such a contraption, but in fact it makes perfect sense for a company billing itself as a full-service environmental company. ENSR's services include air quality management, environmental permitting and compliance, health assessment, asbestos and lead investigation, industrial hygiene, water resource analysis, wastewater management, storm-water management, property transfer assessments, wetlands identification, wildlife management, toxicology services, litigation advice, and hazardous waste management. In essence, the services mirror the nation's environmental laws.

Hanging on a wall, a plaque from Grant Ferrier's *Environmental Business Journal* applauds ENSR for "outstanding achievement" in 2002. On another wall hangs a framed copy of a *Business Week* article on ENSR's management of contaminated sites. There are photographs of the Tongass Narrows in Alaska. There is a framed list of twenty-five-year and twenty-year employees—thirteen of the former, thirty-nine of the latter. There is a photograph of the Rock Beach Hydroelectric Project. A framed color print with trout jumping under a water-fall says "Watershed Preservation." In the company's library, down the hall from Stuart's office, hundreds of reports line the walls, with titles like "Springs and Seeps Assessment, Yakima Training Center," "Yellowstone Pipeline Easement

Renewal Final EIS," and "Velocity Measurements Fish Diversion 6455–730."

While growing population pressures and increasingly strict environmental regulations create a need for environmental consultants, there is seldom an undersupply of people who can—or who claim they can—manage environmental challenges. There are small niche firms, like Mike Rolband's Wetland Studies and Solutions, Inc., and like Lazy Mountain Research in Palmer, Alaska, whose owner and only full-time employee specializes in arctic plants. There are other full-service environmental firms, like Ecology and Environment, Inc., with eight hundred professionals in 27 offices, and SWC Environmental Consultants, with three hundred professionals in 10 offices. There are firms that provide environmental services as part of a broad suite of other technical services, such as URS, with twenty-five thousand employees in more than 300 offices, and CH2MHill, with ten thousand employees in more than 150 offices.

When projects arise, these companies submit proposals, each trying to outdo the other on cost, on expertise, on reputation, on efficiency—in the end generally competing on their ability to fulfill environmental expectations quickly and at the lowest possible cost, whether those expectations are driven by legal obligations, as is usually the case, or by other motivators, as sometimes happens.

"Really good projects," Stuart says, "need two things. They need to be profitable, and they need to be long term. If they're not long term we spend all our time looking for more projects, but if they're not profitable they won't keep us afloat." Without long-term projects and reasonable profit margins, a self-destructive spiral develops in which winning new work constantly diverts the company's efforts. Employees, pressured to maximize billable hours during the normal work week, use their own time to keep the work stream flowing. They spend too much time in the office, too much time away from home, too much time in meetings. They burn out. The environmental consulting industry has a reputation as a mill, capable of grinding up young scientists and engineers, then spitting them out for recycling as employees in less competitive industries.

In the late 1990s, ENSR had some problems, among them an increasingly unmanageable debt load. For a time, there was talk of selling the company to its employees. When this fell through, ENSR was sold from a German conglomerate to Wingate Partners, a venture capital firm capable of covering the company's accumulated debt. The change in ownership was followed by what the *Environmental Business Journal* called "a global campaign to shift the culture towards an employee-centered focus, with an ambitious goal of becoming the 'Employer of Choice' in the environmental industry"; the shift was based on the belief that "engaged employees are productive and loyal, fostering satisfied clients, resulting in revenue growth and thereby satisfying shareholders, who then reinvest in employees." Performance raises and market adjustment raises were put in place. Professional development opportunities were created. For those in a business that sells intellectual services, the message here is simple: Focus on employees and walk like a lion, exploit employees and go the way of the passenger pigeon and the dodo.

Stuart talks about his fieldwork. Although he spends most of his time in meetings and offices, he works hard to be in the field at least one day every couple of weeks. At times, he takes his family with him, as unpaid assistants. His two-year-old son once helped with deer and bat surveys. "Even a two-year-old," he tells me, "can hold a tape measure." Part of this effort comes from a desire to be outdoors, to be a field ecologist, to be with the birds and plants and gophers that originally attracted him to this arena. But there is more to it than that.

"If I spend time in the field," he says, "I know when things make sense and when they don't. If, in a meeting, someone makes an unreasonable claim, I can tell them that they are wrong, wrong, wrong." The extra effort needed to acquire hands-on understanding does not come without cost. He has, on more than one occasion, worked two-day stretches, catnapping and eating when he could. The standard forage for the field ecologist: bags of stale potato chips, cookies, and, when there is time to feast, peanut butter sandwiches.

I tag along with Stuart to Fort Lewis, south of Seattle, where he is working on an environmental impact statement for the army. Stuart sounds apologetic when he explains that this particular environmental impact statement is routine, attracting none of the controversy associated with tough projects like the Crown Jewel and Pend Oreille mines. But, in fact, it is the routine environmental impact statements that are the norm, and understanding this reality is important.

Stuart drives us past guards and a sign that says, "Word of the Month: Pride." Unarmed soldiers in green uniforms mill about, coming and going from what appear to be for the most part desk jobs. But this appearance is deceptive. Fort Lewis was chosen to develop what the army calls "technology-enhanced, fast-deployable and lethal brigades." Soon, the army will be able to drop brigade-strength combat teams anywhere in the world within four days, a full division within five days, and five divisions within a month. They will rely on a new kind of armored vehicle called a Stryker—a light, fast, maneuverable truck named after two war veterans, from separate wars, who shared the same name and who both won, posthumously, the Medal of Honor. The vehicle, a cross between a Bradley fighting vehicle and a very lightly armored tank, hit the ground with a vengeance in Iraq, but at Fort Lewis the concern is with training and developing new brigades. The brigades will count on "just-in-time" supply lines to replace what the army has called "iron mountains" of supplies. There will be changes to equipment, training, and facilities. Eventually, everyone working at Fort Lewis will be affected, and, ultimately, the entire army will change. Chief of Staff General Eric Shinseki has compared this transformation to one that occurred a hundred years ago, after the United States victory in the 1898 Spanish-American War, when the nation suddenly needed an army that could support worldwide ambitions and responsibilities. But there are many differences between 1898 and 2003. One difference is this: Secretary of War Elihu Root, working after the Spanish-American War, could implement a transformation without an environmental impact statement, but General Shinseki, working now, cannot.

"Fort Lewis," Stuart says, "is an oasis in an area that is developing rapidly. All over the country, the army manages lots of land, and Fort Lewis is no exception. Fort Lewis has something like eighty-six thousand acres. We have prairies that used to be pretty common around Seattle before the white man came and stopped burning the land. Now most of the prairies have either been developed or grown over by conifer trees, but on the base fire is part of training and it maintains the prairies. We end up with a lot of species issues. Bald eagles nest on the installation. We have a number of plants that are protected by the Endangered Species Act. Two-thirds of the installation is potential habitat for spotted owls. We haven't seen any spotted owls since the early nineties, but it's potential habitat. The army and the Fish and Wildlife Service agreed to manage for spotted owls even though we haven't found any, so now we monitor for spotted owls. We have endangered western gray squirrels. The pocket gophers out here may be listed under the Endangered Species Act, and there is a butterfly that may be listed. We have gopher and butterfly surveys planned for next year."

In this oasis, over thirty thousand person-days of field training occur every year. Mission Essential Task Lists that drive training include environmental restrictions, and field officers are expected to enforce these restrictions just as they would enforce any other orders. White, yellow, and red stakes mark sensitive areas that are off limits for training, unless you happen to be a gopher or an endangered plant. Wetlands are avoided.

"On certain maps," Stuart says, "there are circles. In the middle of each circle is an eagle nest, and the army can't train inside the circles." Part of the compensation for impacts to eagles includes rebuilding abandoned nests with the hope of helping out eagles. The army hires certified eagle nest restorers. There are three certified eagle nest restorers—the human variety—in the state of Washington.

Inside the base, we pass a Burger King and a Wendys. We pass row after row of base housing. An osprey flies over Stuart's car. We come to a parking lot full of Stryker vehicles—over a hundred of them, in five rows. The army describes the Stryker as "a highly deployable-wheeled armored vehicle that combines firepower, battlefield mobility, survivability and versatility, with reduced logistics requirements." These are brand new and have yet to be fitted with radios, navigation systems, grenade launchers, vision enhancers, and heavy barrel machine guns.

In a warehouse next to the parking lot, several Strykers stand wrapped in plastic, freshly delivered from Toronto and Alabama. A retired army mechanic, now employed by General Dynamics, the company that manufactures the Stryker, gives us an impromptu tour. At our request, he starts one of the vehicles. Though it looks deadly, it sounds no different than a Mack truck. And inside, it smells unmistakably like a new car. There are three benches, a computer mounted in a steel frame, a tiny steering compartment. There is no air-conditioning. "Air-conditioning can be added," the mechanic says. "The commanders' vehicles will have air conditioning. But the infantry loves these things. These things mean they don't have to walk."

In its typical mode of operation, the vehicle will travel across most kinds of terrain at speeds up to sixty miles an hour, carrying a driver, a squad leader, and nine infantry men. The driver training course includes concrete pits with four-foot-wide gaps. It is intended for urban warfare and patrols, with weapon systems designed primarily to provide cover fire for what the mechanic calls "dismounted troops."

"It gets," he says, "about twice the gas mileage of a tank." Clarifying, he tells me that is roughly five miles to the gallon. Each vehicle costs something over a million dollars.

"These won't replace tanks," he says. "This is a whole new machine for a whole new way of fighting. The units will determine what their needs are and how to best use the vehicle. Third Brigade is really helping to define how the Stryker will be used."

While the mechanic talks, Stuart squats and looks under a vehicle, as though sizing up a used car, and before long we have seen all that there is to see. We get back into Stuart's car, an older model four-door blue Cadillac, less fun to drive and less handy in traffic than a fully armed Stryker.

We meet Patrick LaViollette in one of the office buildings on base. Among other things, Patrick manages some aspects of ENSR's contract. "The army develops what we call an IDIQ, an indefinite-delivery–indefinite-quantity contract," Patrick says. While he talks, a military band plays outside. "It's an umbrella contract. We went out and solicited proposals for five years of work and selected ENSR on the basis of a number of criteria. There was more than just price considered. We looked at rates, but we looked at professional qualifications too. From the umbrella contract, we can write a task order for anything that has to do with environment. We can give a task order for an environmental impact statement, but we can also give one for counting gophers and eagles and bears. We can spend up to $750,000 a year."

I ask about Stuart's environmental impact statement. "We'll spend just over a million dollars on it," Patrick tells me.

This is Patrick's second career. Before taking this job, he retired from being what he refers to as "a hard-core civil engineer" with the Department of Transportation. He was building highways when Nixon signed NEPA into law.

"My experience with the NEPA process is more of a traditional approach," he says, "You get a project that you can put your arms around. You know you have to build a highway, and in the environmental impact statement you identify the route that will have the least impacts and you show how to mitigate the impacts. Then you get someone to sign off on it. The environmental impact statement we're working on here is different. We're assessing a process. It's nothing you can get your arms around. When it's done, it's going to be a flexible document. It's going to set thresholds. Instead of saying, 'You can do this, but you can't do that,' it's going to say, 'You can do this until something happens, and then you have to do something else.'"

The bottom line is this: They will monitor impacts, and when impacts reach a certain limit they will move to a new area for training.

Outside, the band grows louder, and before the "Star-Spangled Banner" can drown out our conversation Patrick closes the window. I tell him that what he is describing sounds flexible and less prescriptive than most environmental impact statements. "It sounds good," he says, "but you've got to remember that our primary function here is to train troops. If you start telling the military that you lost a couple of bunnies last week and that now they have to stop training, they're not going to be very receptive."

Under NEPA, the army started with an environmental assessment. They could not, with a straight face, conclude with a "Finding of No Significant Impact," which meant that they could not go right into full training mode with Stryker vehicles. Instead, they limited training to two brigades and put a handful of Strykers on the road while the environmental impact statement was completed. For now, most of the troops ride around in Humvees pretending to be in Strykers. But this is not a big inconvenience. Training is often phased in even if an environmental impact statement does not slow the process. The environmental impact statement should be finished within a year or so, its two-inch thickness ending with a Record of Decision that will finalize the transition from Humvees to Strykers, with all of the accompanying changes in training practices. The issue with the environmental impact statement is not one of whether or not Strykers will be used for training, but rather how they will be used and what level of impact will be accepted before training has to be moved to another site. The army, to the extent that it sees fit, and to the extent to which other agencies have the power to regulate its behavior, will incorporate the suggestions of the environmental impact statement into its training routines.

Patrick took this job to put his daughter—his "semiprecious daughter," he calls her—through college. "I didn't take this job because of a strong commitment to the environment," he says. "I guess I'm as committed as the next guy, but I don't hug trees on a regular basis." He pauses for a moment, then adds, "I guess it would be hard to be against the environment. I watched a guy testify to the state legislature once. The committee was asking engineers, 'Would it be more efficient if we didn't have all these environmental laws?' Really, what they wanted to know was how much all of these environmental laws were costing. Nobody wanted to answer the question. 'Well,' they would say, 'we would still protect the environment. We would still do the right thing.' Why didn't they just answer the question? Why not just admit that they could save maybe 37 percent, or whatever it might be, if they didn't have to worry about the environment? Worrying about the environment costs money."

One of Stuart's colleagues told me that work for the government comprises almost 15 percent of ENSR's workload. They work for the postal service, the Environmental Protection Agency, the Federal Emergency Management Agency, the Forest Service, the Fish and Wildlife Service, the Bureau of Land Management, and all branches of the Department of Defense. Many of these contracts involve NEPA issues. On top of that, many of their private-sector projects stem from NEPA. Although ENSR does not directly track the workload that originates from various environmental statutes, a ballpark estimate of NEPA-derived work for ENSR would be 25 percent. That translates to just under forty million dollars a year for ENSR alone. Looking at ENSR's competition and multiplying through, one starts to realize that real money is at stake here.

I mention to Patrick that some people believe that the NEPA process can lead to real savings by identifying options that might not otherwise have been considered. "I think you'd be hard pressed to find an example of that," he says. "Money is saved through the value engineering process that we do as part of planning, not through NEPA."

I suggest to Stuart that this line of work could lead to cynicism. In response, he tells me about a project he had worked on years before. "I had a situation that just kind of tore you up inside. A guy wanted to develop a bluff on the coastline, but he had an eagle nest on the land. We were able to work something out that left a buffer zone around the nest, a nice win-win situation. He was counting on the money to put his kids through college and for retirement. But there was arbitration, and he was told that he could not do anything with the land. A lawyer made this decision, not a biologist. The owner had other property, though, and when he walked out of the door he said, 'We're cutting down every big tree tomorrow. I won't ever go through this again.'"

Stuart and I, back in his Cadillac, drive around. We look for troops training with Strykers. We have no maps, and the base road system is best described as a labyrinth. We hit dead ends and roads closed for training and live fire exercises. We round a corner and see Mount Rainier, snow covered, in the distance.

"The army does a lot of smoke training," Stuart tells me, "and because Mount Rainier is a National Park and Wilderness Area very strict opacity rules apply." The smoke, intended to provide cover while troops move into position, comes from a mix of diesel fuel and graphite dust. Opacity rules regulate emissions of black smoke; the rules, quite reasonably, do not allow the sun to be blocked by smoke without a special waiver. Stuart goes on to talk about the kinds of impacts that might occur as the army changes its training regimen. Troops will trample habitat. Emissions from vehicles will impact air quality. Dust will be raised. Pygmy rabbits, if any occur near the installations, might be disturbed. There will be noise.

"Lots of the issues that we look at are not that interesting," Stuart says. "Things like routine use of pesticides and low-level noise issues are hard to get excited about. And some of the things we do or promise to do in the NEPA process don't make a lot of sense. A mining company might be asked to spend millions of dollars on wildlife habitat restoration, even though the best they can hope for is something less than pristine. They might be asked to install pipes that trickle water down the side of a mountain, mimicking seeps. Does that make sense? Why not put the money into purchasing pristine land for conservation? And other kinds of development won't face the same scrutiny. A housing development built in the middle of high quality prairie might be required to do nothing more than throw up a few bird feeders."

The army's environmental impact statement, in comparison to the Pend Oreille and Crown Jewel mines', has attracted little attention. While some projects attract tens of thousands of written comments, this project received something like thirty. The public meetings for this project draw no more than twenty or thirty people. Among these are elected officials and neighbors, but also contractors looking for work. Army representatives and Stuart show off posters, pass out brochures, and hand out brownies. The paper tiger named NEPA stalks around, licking its chops. The army has no choice but to stop and think, and this costs money. While the process may lead to cynicism, this fact remains: By design or not, NEPA builds the cost of thinking about the environment right into the project's bottom line. And when one stops and thinks, it sometimes becomes apparent that what one at first considered wise is in fact unacceptable. Sometimes, mines are not built.

CHAPTER FOUR

BIG MUDDY GUMBO

This chapter was written before Hurricane Katrina. As New Orleans is rebuilt, the details surrounding the city's water treatment operations will change, but the fundamental picture will remain the same. Gordon Austin, the focal character of this chapter, survived the storm. A month after the storm, by his own account, he had been "working since the lights came back on sometime in the beginning of September," and he was "beat down like a dog." But because of his efforts and those of his colleagues, water services were back on line relatively quickly for many areas. By October 6, 2005—just over a month after the storm's landfall in Louisiana—the boil-water order was lifted for the city of New Orleans, and by December 8 the boil-water order was lifted for many of the surrounding residential areas.

Just behind Gordon Austin, a sizable chunk of the Mississippi River makes a sharp left detour on its way to the Gulf of Mexico. The city of New Orleans siphons off a hundred million gallons a day, sucking it in through six oversized soda straws, the smallest more than big enough to swim through. As the water flows downhill from the river to the city, mud will be removed and chlorine will be added. The cleansed river will wash cars, water lawns, slake thirsts, bathe sweaty bodies, flush toilets, extinguish fires, and dilute cheap liquor on Bourbon Street. After use, it will be cleaned again, and then sent back to the river, little the worse for wear.

As a fifty-four-year-old employee of the New Orleans Sewerage and Water Board, now into the third decade of his career, Gordon understands this process. When he started work, soon after returning from duty in Vietnam, he was one of two employees in the organization's environmental group. In those days, one aspect of his duties involved checking grease traps at some of the city's finest restaurants and at some of its worst restaurants. Since then, he has watched the Safe Drinking Water Act and the Clean Water Act evolve. He has shown professors how to mine data from his employer's century-old archives. In 1982, at the request of the then adolescent Environmental Protection Agency, he wrote, and then rewrote, the first pretreatment plan for New Orleans water. Not unfittingly, he occasionally teaches courses on the history of water treatment. Today, he heads the board's Environmental Affairs Group, answering directly to a deputy director, who answers to a director, who answers to thirteen board members, including the city's mayor. And, if anything goes wrong, the mayor answers to the voters.

"But the voters," Gordon tells me, "only notice if the water turns brown. And New Orleans water doesn't turn brown. People in New Orleans are used to high quality water. If they want brown water, they go to Atlanta. Here, they are so used to clean water that most of them don't even know where it comes from. Ask them where the water comes from, and they'll say, 'It's from the tap.'"

Gordon sports a graying mustache and shaded sunglasses. His years with the Sewerage and Water Board have seen his hairline paddle upstream, though it has a long way to go before it reaches Baton Rouge. Framed against the river, leaning over a rail to snap photographs of the water intake, Gordon sports a New Orleans profile that roughly approximates the deep bend of the river as it flows around the city. This profile reflects the fact that he is, in his own words, a glutton. His choice of the word—"glutton," rather than "epicure" or "gourmet"—bespeaks a frankness and a self-effacing humor not expected of someone sent to talk to an environmental writer about New Orleans water, especially now, just days after what appears to be at least a temporary cease-fire in the long battle over water system privatization that, among other things, sent one board member to federal prison. His photography, too, says something of his character: He clicks away at intakes he has seen a hundred times.

"See the spray?" he asks, in an accent that exists only in the strip of land south of Lake Pontchartrain and north of the Gulf. "The spray blocks out floatables. And the booms keep the area clear, too." Floatables are logs and Styrofoam coffee cups and empty plastic milk bottles and pieces of rope and dead birds and leaves and half-filled beer cans and oily bilge water illegally pumped from barges. Twelve jets, resembling pressure washers, blast the surface. The booms—algae covered yellow plastic tubes that hold curtains extending several feet below the surface—further defend the perimeter around the intake pipes.

A tanker heads upstream between us and the river's far bank, a mile away; it fights the current the same way that Mark Twain's riverboat would have fought the current, its engines straining to make headway and kicking up a surfable wake. Coming the other way, headed toward the Gulf, a passenger barge, painted blue and white in a low-end attempt to remind paying customers of a riverboat, glides with startling velocity, empty, perhaps headed to pick up tourists somewhere near Jackson Square. Behind the boats, a grain elevator and a power plant nest side by side on the river's west bank. The grain elevator is famous for having once exploded, some time in the 1980s, Gordon thinks. "There's some fat catfish over there," he assures me.

Below us, in the realm of fat catfish, water rushes silently from the river to the heart of New Orleans. Other body-part metaphors apply. The drinking-water purification plant is the stomach, digesting the water from the river to separate the good stuff from the bad. Sixteen hundred miles of water mains serve as arteries. Almost fifteen hundred miles of sewer pipes are the veins. And the wastewater treatment plant on the east side of New Orleans is the kidney. Leaving the body metaphor behind, the Mississippi River downstream from New Orleans is the toilet. All of which works out reasonably well as long as we ignore the more than eighteen million people living upstream who also use the river as a toilet, who consciously or unconsciously use the river as their own special way of saying hello to New Orleans.

The Mississippi has been ranked number one as the nation's most polluted river. It has been said—incorrectly, but it makes a point—that every drop of Mississippi River water has passed through four water treatment plants by the time it reaches the Gulf of Mexico, ninety miles downstream from New Orleans.

Flush a toilet in Minnesota and four weeks later the flush passes in front of the Audubon Zoo. Open an outlet valve in a Missouri paper mill and two weeks later the effluent slides past the French Quarter. Spray your soy beans in Arkansas, wait for the rain, and five days later pesticide residues slip underneath the Huey P. Long Bridge. By the time the river reaches New Orleans, it carries fecal coliforms, nitrate and phosphate fertilizers, linear alkylbenzenesulfonates, dioxin, polyethylene glycol, halogens, ethylenediaminetetraacetic acid, an antiphlogistic drug called naproxen, estrogen, hexachlorobenzene, caffeine, various volatile organic compounds, atrazine, PCBs, lead, and tris-2-chloroethylphosphate. And other stuff. Between 1990 and 1994, more than seven hundred million pounds of toxins ran into the river, counting only those reported to the government. Well within the bounds of acceptable hyperbole, the Mississippi River could be called the biggest sewer in North America.

On the other hand, the Mississippi River carries lots of water; if pollution were scotch, the river would be substantially weaker than near-beer. To those whose image of rivers come from trout streams or white-water rafting trips or East Coast trickles like the Potomac or the Delaware or the Hudson, the Mississippi does not look like a river at all. At normal flows, the Mississippi River in New Orleans is over a mile wide, funneling three hundred billion gallons a day from thirty-one states and two Canadian provinces. At flood stage, if the levees give way, the opposite bank ducks below the horizon. The river flows through New Orleans in a natural channel reaching close to two hundred feet deep. Dead trees tumble along the bottom, occasionally hitting updrafts that send them shooting to the surface. The Corps of Engineers lines certain areas with massive concrete mats in an effort to control the river, but the mats disappear under sand and mud or simply wash downstream. Surface boils and ephemeral whirlpools and back eddies along the shore bring texture to the river's surface, and 210 million tons of sediment per day bring texture to the water itself. The river's name came from the Chippewa language: Mississippi, meaning "the great water" or "the gathering of waters." It has also been called the Big Muddy.

For a moment, Gordon and I look at brand-new video cameras that monitor the surface above the water intake. They are odd for two reasons. First, they are brand new, while everything else out here carries the relaxed New Orleans patina of deferred and forgotten maintenance. And second, the cameras are not connected. "They're security cameras," Gordon tells me. "But there's no real issue out here for security. We went through this in the seventies, too. People used to worry about someone dumping LSD in the water supply, stuff like that, in the seventies. But you'd need a trainload to do any damage. The same thing is true today. The water intakes just aren't a security risk. There's too much water going by."

Since the inception of the Clean Water Act in 1972, Americans have spent more than eighty billion dollars cleaning up rivers and lakes. Since the inception of the Safe Drinking Water Act in 1974, they have spent an additional sixty billion dollars to improve drinking water quality at the tap. Rivers no longer burn. Americans seldom contract diseases from their taps. But this does not mean that everything is perfect. This does not mean that a responsible person, such as, say, Darek Malek-Wiley, president of the Mississippi River Basin Alliance, could not stand on the bank of the river, just downstream from where Gordon and I now stand, and publicly state, with a straight face, "What we're getting here is a gumbo of different kinds of chemicals."

Before going on, understand that Gordon does not unquestionably love and respect his employer, and understand that his impending retirement frees him to speak his mind. He has told me, among other things, that one of the Sewerage and Water Board's major repair programs is "about as useful as taping hundred-dollar bills to a sewer pipe." He has said that the prison sentence recently awarded to a board member was not only deserved, it was too lenient. He has explained that a six hundred million dollar consent decree ordered by the Environmental

Protection Agency was not necessarily a bad thing for the city. But when I ask him about New Orleans water, he tells me that everything is in good shape. Despite population increases along the river, despite thousands of new chemicals that find their way downstream, despite chemical spills, despite agriculture's penchant for dumping pesticides onto the land only to see them wash into the river, despite everything, Gordon assures me that water flowing to the taps of this city, his city, is good water.

I ask him the obvious question: "Do you drink the tap water?"

"Oh yeah," he says. "All the time." And he clicks another picture of the surface above the water intake.

In a blue Lumina with the words "New Orleans Sewerage and Water Board" stenciled on the door, we drive around the corner and down the block to the Carrolton Water Purification Plant, a collection of buildings and pipes and tanks dating from the 1920s, ranging in appearance from art deco to industrial drab, from walls of white stucco and windows with red trim to rust-stained tanks made from riveted steel plate. Weeds grow through cracked concrete.

By comparison to the process used here, the buildings are as new as morning dew. "Water treatment hasn't changed a lot," Gordon says. "I'm not talking just about here. I'm talking everywhere. Most of this stuff goes back to the Babylonians."

Six-thousand-year-old written records describe attempts to improve the appearance and taste of water. Four-thousand-year-old Hindu texts direct readers to boil water. The practices of charcoal filtration and of allowing sediment to settle out in stilling basins both predate the Greeks. The Egyptians used alum to help pull particles out of the water. In 1627, Sir Francis Bacon published information on water purification through percolation, filtration, boiling, distillation, and coagulation. But the practices known since antiquity were not widely used. Although Anton van Leeuwenhoek published sketches of protozoans swimming in water a century and a half earlier, it was not until 1855 that London's Broad Street Pump Affair led to the realization that water can transmit disease. This realization came from an epidemic of cholera, a disease that turns skin black and blue through dehydration and puckers hands and feet, sometimes killing victims within twelve hours. The nineteenth-century physician John Snow traced London's 1855 epidemic to the Broad Street Pump and realized that the water tapped by the pump was contaminated with sewage. He surmised that people were contracting cholera by drinking dirty water. It was soon known that dirty water was alive with pathogens. Dysentery and typhoid swam with cholera. By the late 1800s, people were experimenting with ozone, now commonly used to disinfect public water supplies in Europe. Chlorine was used for the first time in 1908.

Gordon parks near two railroad cars, one white and one black. Chlorine, apparently coming from a small leak or a vent in one of the cars, bites my lungs. If either of these cars released its full load to the atmosphere, people downwind would die, but adding it to water in the right doses saves millions of lives. The right dose, for New Orleans, is about three tons each day. I follow Gordon past the railroad cars and up a flight of stairs to stroll along the top of what he calls "the primary settling basins." The basins—long pools that stretch out into the distance—reflect the sun. Just here, near the top of the steps, the water does not look any more inviting than the Mississippi River itself, but as it moves through the basins it drops its load of sediment. I walk behind Gordon, following him along catwalks. Gulls with black heads, perched around the basins, look at us and screech, their calls mixing with the sound of falling water and an electric pump. The addition of chemicals—a ton a day of polyelectrolytes and twenty tons a day of ferric sulfate—help pull out the sediment. The chemicals help convert suspended sediment into what Gordon calls "sludge." I like the word; it is at least subtly onomatopoeic. As the sludge becomes more concentrated, it takes on the appearance of wet concrete, a limey tan paste.

"This is where some of the research for the Safe Drinking Water Act was done," Gordon says, "right here at this plant." He turns to talk as he moves along the catwalk, and he speaks above the noise of falling water and an electric pump and screeching gulls. "You know about trichloromethanes? Trichloromethanes form when you add chlorine to the water. They're a cancer precursor. The chlorine killed the pathogens but turned them into something that could cause cancer. So we still add chlorine to kill the bugs, but then we add ammonia. We call it chloramine disinfection. This research was all going on about when I started work, right here. I didn't know much about it at the time, but it was all going on."

He stops for a moment and we stare out over the water. "People thought we had high cancer rates around here, that the river was a bad source of water and that it caused cancer. But the data were skewed. Lots of people at Oschner died from cancer, but that's because they're coming here from all over the Caribbean. People don't know this, but New Orleans is really a magnet for people in the Caribbean. When they get sick they come to New Orleans. They check into Oschner. So that made it look like our cancer rates were higher than normal. But it gave New Orleans a chance to be part of this research."

Moving downstream along the settling basins, clearer water flows over concrete sills into other settling basin cells. A lime scale covers the edges of the sills, giving them the look of flowing rock that one might see in a limestone cavern. The water now has a pleasant green tint. I can see bottom. It is warm enough, I think, for a swim, though I suspect this would impress neither Gordon nor his customers.

"We're lucky," Gordon says. "When the sediment comes out, it brings lots of other stuff with it: pesticides, waste from chemical plants, that kind of stuff. A lot of this stuff doesn't need special treatment. I call it incidental treatment. We're just lucky, really."

The luck extends beyond incidental treatment to treatment itself, any treatment. The level of treatment we routinely expect in the developed world, the level of treatment that we consider a necessity, is, for most of the world, an unaffordable luxury. But just because water is not clean does not mean that people will not drink. A 1 percent deficiency of water makes us thirsty, a 5 percent deficiency gives us a fever, at 8 percent we stop making saliva and our skin turns blue, at 12 percent we die. So people drink, and globally contaminated water kills more than ten thousand people each day.

From the director of the Department of Women's Affairs in Namibia: "Our priorities are different from those of the current American feminist movement. Women in America are shopping for dishwashers; women in the Third World are looking for water to wash the dishes."

From a rural father in Senegal whose daughter left for the city: "My daughter got tired of carrying water."

From a coordinator of the National Conservation Strategy in Pakistan: "I know of nine-year-old farm lads in jail for murder over water disputes."

The United States is not as distantly removed from water problems as one might assume. New Orleans enjoyed cholera epidemics in 1832, 1849, and 1850. Before 1905, the city relied on cesspits for sewage; there were sixty-seven thousand of them in 1899. Drinking water came from cisterns fed from rooftops and from casks of river water that served as settling ponds, allowing river mud to settle out. Federal regulations for drinking water appeared in 1914, but they only applied to water used in interstate commerce. In 1972, just a few years before Gordon started work, an analysis turned up thirty-six unwanted chemicals in New Orleans drinking water. The Safe Drinking Water Act as we know it today did not appear on the books until 1974.

Israel's Shimon Perez wrote, "If you want to save your children from poverty, pay attention to water." And the United States pays attention to water. The Safe Drinking Water Act, intended to protect public health, governs something like two hundred thousand water systems, some publicly owned and some privately owned. The act is complex on its own, but making it even more complex are the interweaving of state and federal requirements and Congress's inability to resist opportunities for improvements. At the end of the day, though, the act is simple: By the time water reaches the tap, it has to meet certain standards. Barium cannot occur at more than two parts per million. Mercury cannot occur at more than two parts per billion. For lead, it is fifteen parts per billion. For ethylene dibromide, fifty parts per trillion. To gain perspective: A part per million corresponds to one penny out of ten thousand dollars and a part per trillion corresponds to one penny out of ten billion dollars. Although the allowed levels may at first glance seem to approach the ludicrously low, they are not arbitrary. Someone has worked out, for example, that more than two parts per million of barium can increase blood pressure.

More than two parts per billion of mercury damages kidneys. Any amount at all of the pesticide ethylene dibromide can cause liver problems, stomach problems, reproductive problems, and cancer, so even the allowed fifty parts per trillion may be too high.

The Safe Drinking Water Act passed as an unfunded mandate, forcing the costs onto state and local governments. A government report issued in 1995 argues that all the act really does is standardize expectations across the nation. Under the act, water in Buffalo will meet the same requirements as water in Des Moines, which will meet the same requirements as water in New Orleans. The report says that water systems spend about two billion dollars a year to comply with Safe Drinking Water Act requirements that go beyond preexisting regional requirements. A typical early 1990s household would pay about four hundred dollars per year for water, but of this only about twenty dollars was needed to meet new requirements imposed by the act. The act had costs, but it brought benefits, too, and the government estimated the value of these benefits. Example: Avoiding a case of cancer by regulating ethylene dibromide costs five hundred thousand dollars—a bargain, especially if the cancer case avoided is your son's, or your daughter's, or your mother's. Regulating the pesticides atrazine and alachlor, on the other hand, could be more pricey, on the order of several billion dollars per cancer case—more, according to the report, than the five million or so dollars that the government considers to be the value of statistically avoided deaths. But still a bargain if the cancer case avoided is your own.

"Being in this environmental field," Gordon tells me, "is kind of intimidating to a lot of people, to a lot of engineers even, but especially to administrative people. I don't do it, but it can be sort of a game, kind of funny. You could always use the

Environmental Protection Agency to scare people, because of fines and jail time." If the obvious value of clean water is not seductive enough, avoiding fines and staying out of jail could reinforce the will to comply.

While we drive from the settling basins to the rapid sand filters, Gordon talks in general terms about the evolution of water treatment. "Now what we do," he says, "is fine tuning. Heavy metals have been done to death. Pharmaceuticals and personal health products are the next step. They may be a real problem. We don't know." Estrogen and the anti-inflammatory naproxen show up in drinking water. The levels are low, the effects uncertain, but the stuff is there, flushed into toilets but finding its way into the drinking water supply from somewhere upstream. "We know," Gordon says, "that there's a lot of Prozac out there." The knowledge that we are drinking slightly used estrogen and naproxen may be depressing, but the Prozac should help us cope.

Out of the car, Gordon leads me along walkways over the rapid sand filters. Gulls screech overhead. Water from the settling basins flows here, looking almost drinkable. It is already chlorinated. The chlorine's cancerous effects have been mitigated by ammonia. As a bonus, a touch of fluoride has been added to prevent tooth decay. The water moves downward through layers of sand and gravel and anthracite coal. "What we see here," Gordon says, "this is not finished water, but what comes out underneath is." We look through the water at the sand. Forty-four filters can handle more than two hundred million gallons a day, all pulled through the filters by gravity.

We walk from bright sunlight through an art deco doorway into the filter gallery. The noise of flowing water and machinery and drips, all of it with subtle echoes, muffles our voices. Pipes run everywhere. Among the pipes, a hollow cedar log, fifteen feet long and a foot in diameter, hangs from a ceiling with a sign that says, "Style of water main used in New Orleans previous to 1925." This makes the log only slightly older than the filter gallery itself.

We walk down steps into what might be thought of as the building's cellar. "All the pipes we have here," Gordon says, "come off the bottom of the sand filters." Clean water, fit for drinking, drips from pipes.

Back upstairs, we look at gauges and valves mounted on panels that must have been built about the same time that engineers touted vacuum tubes as the latest innovation in electronics. "What I love about this place," Gordon says, "is all this Frankenstein-era stuff. But it does the trick." We look at hollow glass balls that are the size and shape of the crystal balls used to bilk drunken tourists in the French Quarter. Finished water bubbles up into the glass balls, where it can be inspected one last time.

"If it is muddy in there," Gordon says, "you know you got a problem."

But it is not muddy. The stuff that gives the Mississippi River a look of latte is gone. The water inside the glass bulbs could pass for gin; in some bars on Bourbon Street, it probably does.

And that is it. The water is done, converted from liquid mud to drinking water, from latte to gin, first by settling, then by filtering. In the filter gallery, we take turns drinking from a water fountain.

"This," Gordon says, "is as close to the source as you will get. This is good water."

Before leaving the grounds we stop in an office building to talk to Marvin Morgan, one of Gordon's long-term colleagues.

A leadership award and a special recognition award hang on the wall. More interestingly, a collection of bottled water stands on top of a filing cabinet. This is not just any bottled water, though; these are collector's items from events like Mayor Ray Nagin's inauguration and a water convention. A consulting company sent him a sample. The city of Denver, with mountains in the background of its label, sent him a bottle.

"A lot of the bottled waters come from the public water system," Marvin says. "They'll filter it or purify it to give it a uniform taste. Sometimes they'll distill it down to pure water and add salts. But it comes from the public water supply."

Several of the bottles contain water from the lower end of the Mississippi River, cleaned and processed right here on the grounds, then bottled and labeled, classed up, like a longshoreman dressed for church.

Almost inevitably, conversations with New Orleans Sewerage and Water Board employees turn to pride in their work. "We have kept up," Marvin says. "We've been doing about the same thing for 105 years, but there have been changes. Many times, changes have been psychological and process-control oriented. But we've been keeping up. Studies show that people need new plants to stay in compliance with new regulations. They build the plants and what do they get? The same water they got from the old plants. But process optimization is very successful, even with the old plants. And that's what we are attempting here. We have modernized our feeds—we've gone to more precise chemical feeds. We've changed some addition points. We're flash mixing more efficiently. We just changed our addition point for lime, and that's going great. We refurbished our secondary settling basins a few years ago. But the process really hasn't changed."

I ask about the huge investment needed to maintain and upgrade existing water systems, estimated at a cost of about a trillion dollars within two decades. Cost of needed repairs in New Orleans alone has been estimated at over a billion dollars.

"There are very real things there," Marvin says. "Everybody in the nation—all the major cities were built about the same time. In the late 1800s, early 1900s, with the advent of the germ theory of disease and the discovery of chlorine, people found that adding chlorate of lime made typhoid rates drop. They didn't understand why right away. But all those things came together at the same time. Now we hear people saying that all we have to do is get rid of chlorine and go to ultraviolet treatment of water and use membrane filtration to get rid of cancer-causing agents—simplistic, horribly simplistic. This is like hippie groups in the sixties moving to the hills to live off the land. And as they're starving they come to a realization that you need an infrastructure. What we need is new settling basins, new pipes."

In a corner of Marvin's office, there is an antiprivatization sign, two feet long and a foot tall, with white letters on a blue background: "Do it right Sewerage and Water Board. NO to All Bids! Local 100 SEIU. ACORN. Public Citizen." The Sewerage and Water Board workers have just won another major battle against those who wanted to privatize their organization—those who, as Marvin sees it, wanted to sell off their jobs and their loyalty and their commitment to bringing clean water to New Orleans; those who, as Marvin sees it, were more attracted to the greening of their own bottom lines than to delivery of high quality water.

But privatization of water services is nothing new. A private water company was operating in Rhode Island in 1772, and the Manhattan Company was servicing New York City by 1799.

Today, the private sector services about 19 percent of the market in the United States. What is new, though, is the aggressiveness behind privatization. As claims of waste in government become increasingly fashionable, claims that privatization will end millions of dollars of government waste become increasingly common. Now, the claims are more common than gulls at a water treatment plant. The prize for the private sector is huge; the contract for New Orleans could bring in more than a billion dollars over twenty years. *Fortune* magazine has described water as "one of the world's great business opportunities" and said that water "promises to be to the twenty-first century what oil was to the twentieth." Between 1994 and 1998 there were 139 water deals worth over a billion dollars. Now, private-sector management of public water supplies brings in two hundred billion dollars a year. The World Bank expects water to become a trillion-dollar-a-year industry within two decades. These are very seductive numbers.

Multinational companies run water systems for 7 percent of the world's population. The two largest multinationals, Suez and Vivendi, both French and both operating under enough different names to fill a small phone directory, supply water to more than two hundred million people. The multinationals cannot ignore the United States, where most people drink water from government-run water systems. In 1999, this interest drove Vivendi to buy what was then the largest private water company in the United States, U.S. Filter, for more than six billion dollars. The acquisition positioned Vivendi with seven hundred offices, plants, and factories, along with fifteen thousand employees, scattered through Canada and the United States. Overnight, they took control of the water supply for almost fourteen million people in more than six hundred communities. They became the behind-the-scenes owners, twice and thrice removed, of several competing brands of bottled water, including Culligan, which shows up in more than three million North American households. The acquisition also gave Vivendi, operating under the U.S. Filter name, the first-glance appearance of being an American company, a not altogether bad feature for a company trying to tap into and expand its grip on the North American market.

Experiences with privatization have not been universally favorable. When water was privatized in an Atlanta suburb, the Environmental Protection Agency had to issue a notice calling for use of boiled water. Not too long after that, brown water flowed from the taps. "I had parents calling in tears," said Gordon Certain, president of the area's civic association. "The things that have happened to water here have sure scared the hell of out a lot of people."

Overseas, where aggressive privatization has been building for years, stories of problems have accumulated like sludge in the bottom of settling tanks. In England, privatization, touted as a cost saver, was accompanied by a 50 percent price increase within ten years, prompting the *Daily Mail* to call water "the biggest rip-off in Britain" and "the greatest act of licensed robbery in our history." In Brazil, Panama, Peru, Columbia, India, Pakistan, Hungary, and South Africa, privatization bred protests. In Paraguay, protests grew rowdy and were quenched with water cannons. In Bolivia, when privatization drove water prices up 35 percent, forcing some households to spend up to a fifth of their income on water, riots broke out. Six people were killed and officials of the company running the system were told that their safety could not be guaranteed.

On the other hand, thousands of systems are privately managed, but the same few dozen appear repeatedly in what has been called "the emotional rhetoric" of antiprivatization. Troy Henry, a regional manager of the Suez-owned subsidiary United Water, once worked for IBM. He wants to see the private sector do for water what IBM did for computing. He believes, among other things, that the private sector is better able to retain "the best and the brightest and most talented people." A U.S. Filter corporate statement explains that the company was created by investors who "recognized the urgent need to better manage the business of water." These investors believed that a larger company would be capable of providing a full suite of services in an industry that "traditionally has been fragmented, populated by small companies with niche specialties." The World Bank and the International Monetary Fund see merit in privatization, too; in 2001, the World Bank approved loans worth $554 million for drinking water and wastewater treatment, of which just over half required privatization. Debra Coy, vice president of Schwab Capital Market's Washington Research Group, puts it simply: "Running water systems is a business." Gerald Payen of Suez agrees. "We purify water and bring water to your home," he says. "We provide a service. It has a cost, and somebody has to pay for it."

Peter Gleick, president of the Pacific Institute of Studies in Development, Environment, and Security, takes exception to the idea that privatization is a good thing. "These are long-term monopoly contacts," he says. "This isn't free enterprise or a competitive market."

Occasionally, bribery of public officials has been a tool used to gain market share, to secure long-term monopoly contracts. In New Orleans, Katherine Maraldo, a New Orleans Sewerage and Water Board member, was sent to prison for accepting something like eighty thousand dollars in exchange for her support of a contract extension. Maraldo's response, upon sentencing: "Your honor, I too have spent a tremendous amount of time wondering why I did what I did."

The business case against privatization has not avoided notice. In South Africa, more than a hundred thousand cases of cholera have been blamed on privatization efforts; when customers failed to pay, their water was cut off, leaving them to drink free water from rivers and streams even though they knew it might be contaminated. "The cutoffs," according to David Hemson of the Human Resources Research Council, "cost them more money in the end in dealing with the disease that resulted from it." What seemed a good deal, what seemed to be sound economic policy, what followed the rules of capitalistic fair play, turned out in South Africa to be not only inhuman but also a fool's bargain, at least as foolish as burning coal in preference to letting wind spin a turbine. A requirement to stop and think might have saved both lives and money.

There are examples of government-run water providers demonstrating that they can compete with and out perform the private sector. Through cross-training, staff reduction via attrition, improved procurement practices—through all the approaches typical of the business world and, supposedly, atypical of the public sector—government providers in Houston, Colorado Springs, and San Diego have cut the costs of operations by 20 percent and more. New Orleans, too, through the employees of the Sewerage and Water Board, has demonstrated its ability to reduce costs: In the competition to win the right to run the city's water system, the lowest bid came from the employee bid team.

"Marvin," Gordon says, "was part of our employee bid team. This was formed from the union. And they got involved with a local socially active group, ACORN, and with a national group, Public Citizen."

"The current privatization," Marvin says, "is over. But it's interesting that the mayor has not formally ended it by act of the board. We believe the vultures are still circling. It may not be the privatization envisioned four years ago with United Water that devolved into the mayor's association with U.S. Filter, but it may take on other forms."

"It's too lucrative an opportunity to ignore," Gordon says.

On the way out, Gordon and I look at pictures posted in the hall. Black-and-white photographs from the 1920s show pipes big enough to walk through being dug into the ground. Portraits of water managers date back to 1904. Gordon studies them, looking closely at the faces, then the dates. It is as though he is looking at a high school album or pictures of the team. He is on the team. Privatization, for Gordon, would be like being thrown off the team. Worse, it would be like firing the whole team and replacing it with players more interested in a bottom line than in the game itself, the game of clean water for the community.

We stop at the purification plant's gas pumps and talk to Norden Mayfield Jr., a truck maintenance worker at the plant, who helps Gordon fuel the car. When Gordon tells him I am writing a book, Norden wants to be sure I have his name right, and he spells it out for me, then spells it again. "You put this in your book," he says. "Nobody do it better than New Orleans. All that stuff that flows from up North comes to us and we clean it out. We at the end of the river but we drink it. Nobody do it better than New Orleans."

Like Death Valley, New Orleans sits below sea level, but New Orleans is maintained by an integrated system of levees, drains, and pumps. We stop at the end of the Seventeenth Street Canal, a giant storm-water drain for New Orleans and Jefferson Parish. The canal flows through giant steel grates into a pumping station. From there, ten thousand cubic feet a second can be pumped uphill into Lake Pontchartrain. The lake connects, unimpeded, with the Gulf of Mexico.

For people accustomed to cities built on high ground, accepting the cross section of New Orleans can be challenging. In cross section the city looks like a lopsided bowl. On one side, the levees along the Mississippi River stand twenty-five feet above sea level; on the other side, the banks of Lake Pontchartrain stand ten feet above sea level. In the middle, in the heart of the city, the ground is, for the most part, a foot or so below sea level. Near the bottom of the bowl, it dips to five feet below sea level. Roofs of some homes stand eyeball to eyeball with sea level. Roofs of other homes must look up to find sea level. Boats on the Mississippi River look down on the city.

Even in New Orleans, water will not, of its own accord, flow uphill. Hence, there are the levees and pumps. Breach the levees or shut down the pumps, and the city becomes a swimming pool. The system can handle an inch of rain in the first hour and a half inch every hour after that; when more rain falls, feet get wet. This is not a question of if, nor even when, but rather one of how often; rain falls often exceed pumping capacity.

"The New Orleans Sewerage and Water Board," Gordon says, "is unique in dealing with drinking water treatment, stormwater management, and wastewater treatment, all within the same organization."

Steel screens, traps for trash captured by water flowing through New Orleans, stretch before us, the length of a football field. A concrete road, wide enough for garbage trucks, runs alongside the screens. Big red tracks hold automatic rakes that sweep trash upwards over the screens, dumping screened trash onto the road. When it rains, trash piles up quickly and trucks wait in line to haul it away. Right now, two days since the last rain, the water and its trash flow at a more leisurely pace. A potato chip bag, empty black plastic oil bottles, basketballs, Styrofoam cups, a plastic toy hand grenade, shrub trimmings, and a tire float against the screens. Empty water bottles bob shoulder-to-shoulder with the other trash.

"What you see there," Gordon says, "is a whole lot of convenience."

Graffiti decorates the concrete sluiceways adjacent to the pump station: "Gangsta was here" and "POW." A snowy egret and a blue heron stand next to the running water, next to the graffiti. Screened water is pushed uphill by pumps before flowing through three miles of canals to a discharge point into Lake Pontchartrain. The water—rainwater that has rinsed the streets and roofs of New Orleans—enters the lake without further treatment.

"The Corps of Engineers wanted to look at treatment for this water," Gordon says. "They did all the math, wrote a thick report. What they figured out is that they would need to convert half of Lake Pontchartrain into a settling basin to handle peak flows."

On the downstream side of the pumps, having slipped through the screens and survived a wild ride through pump impellers, a plastic glove floats to the surface, reaching upward like a bloated pale hand waving good-bye as it heads to Lake Pontchartrain.

We drive past vacant lots and abandoned cars to reach the East Side Treatment Plant. The smell of hydrogen sulfide, emanating from sewage, greets us. Close to where we park, under a set of stacks that belch steam, we feel a mist.

"Tell me," I say, "that it's just distilled water coming off the stacks."

"I sure hope so," replies Gordon, leading me from under the stacks to stairs that lead to a maze of sluiceways. They look similar to those we saw earlier, at the Carrolton Water Purification Plant. But these are different. These are influent channels and sluices for sewage water. This is where the metropolis flushes the john. Used water drains to right here. I brace myself. I remind myself, not for the last time during this visit, that a hundred years ago the city relied on sixty-seven thousand cesspits. I remind myself, too, that cesspits were an improvement over sewage in the streets, a reality for most cities into the 1800s. I remind myself that in a less aseptic culture, where common sense and daily needs supercede prissiness, this stuff is nothing less than valuable fertilizer. And then I have a look.

Brown water flows past. The brown is deeper than that of the Mississippi River. Tan foam froths here and there. If the Mississippi River is latte, this is espresso. There are no floatables. In fact, this is, by and large, just dark water. Each flush sends three to five gallons here. On average, flushing uses 26 percent of the water sent to homes—water from places like the Carrolton Water Purification Plant, water fit to be bottled and sold at exorbitant prices, reduced to this with one easy flush.

Not too long ago, water from here would have gone straight to the river. The Clean Water Act changed all of that. Nixon, believing that the Clean Water Act was too expensive, vetoed. Congress, thinking enough is enough, overrode his decision. The nation, ambitious as always, wanted surface waters that would be "swimmable and fishable." People were fed up with rivers like the Cuyahoga bursting into flames. They did not like reports like the one released in 1969, claiming that three million dollars worth of fisheries were destroyed each year in the Chesapeake Bay because of poor water quality. They did not like the smell emanating from the Buffalo River. They did not like the fact that two-thirds of the nation's lakes and rivers were not considered safe for swimming. They did not want to know that bacteria counts in the Hudson River exceeded safe levels by a factor of 170. Somehow, Americans could not accept the idea that their bass and pike and catfish were breathing toilet water.

The 1972 Clean Water Act, passed when sewage was routinely dumped without treatment into lakes and rivers, called for "zero discharge of pollutants into navigable waters by 1985, and fishable and swimmable waters by 1983." Since then, something over eighty billion dollars in federal grant money has been spent, much of it matched by state and local tax dollars. And while not perfect, it has, by and large, worked. Among other things, it has led to construction of facilities like this one—filters that have to be passed before water is discharged into the river. While its zero-discharge goal may not have been met, while something like 39 percent of rivers and 45 percent of lakes remain polluted, cities no longer flush their collective toilets straight into rivers and lakes. At the Carrolton Water Purification Plant, in spite of flow rates nearly doubling as the population of New Orleans grew, the effectiveness of treatment improved fivefold between the early 1970s and the early 1990s.

Gordon and I walk next to the sluiceways. A tide line of brown scum and the occasional condom lines the edge of the walkway, evidence of recent high water levels. Even now, the sluiceways have only a few inches of freeboard.

"You can't swim in that water," Gordon informs me, apparently concerned that I might want to take a dip. "It's so aerated that you won't stay afloat." Life rings hang on handrails, just in case.

From the sluiceways, we move to the reactors. They are sealed concrete pools into which pure oxygen is pumped. The process here is all about bacteria. The bacteria—at least the ones that do the job most efficiently—need oxygen. They metabolize the sewage, converting it into more bacteria, using oxygen in the process. When they die, they are removed. They become a component of sludge—a different sludge than that of the Carrolton Water Purification Plant, but sludge nonetheless.

"We have good sludge," Gordon tells me. Juxtaposed, the word "good" next to "sludge," oxymoronic, they somehow fail to ring. But Gordon is serious. "One of my dreams," he says, "is to use the sludge. We have all these marshes around here sinking and turning into open water. They need sediment. We could fill them with sludge."

From where we are standing, we can see open water and scattered dead cypress trees, a common scene in Louisiana, where the ground is sinking and land is becoming part of the Gulf of Mexico. "See all these dead cypress trees?" Gordon asks. "And see those live ones over there? The live ones are all growing on sludge. We used to put the sludge there, but we can't anymore. Now we burn it and send it to landfills. What a waste."

I think for a moment about Mike Rolband and his wetlands, almost inconceivable just here, on top of concrete tanks full of sewage. Below us, in the depths of the concrete cauldron on which we stand, pure oxygen bubbles through sewage.

The heat, the long day of my questions, the sewage—Gordon complains that he is not feeling well. This surprises me. He has the energy and cheerfulness and enthusiasm of someone feeling much better than not well.

"Yesterday," Gordon says, "I had a proposal due at five o'clock. It was one of these things that needs ten copies and an electronic copy and a cover letter and all the rest of it. At twelve o'clock I get a call from this guy who says he wants to put his name on the proposal and add a few things to it. Well, I'm trying to get out of the office by one. I had plans. Where has this guy been for the past three weeks? When the last mayor left office, he left a hundred-thousand-dollar grant sitting on the table. Well, the grant he had wasn't worth doing. It was a good idea, but it had already been done. So I wanted to put together something that would do some good instead of just spending a hundred thousand dollars. One of the ideas was to pay a consultant a hundred thousand dollars just to do a little review of what kind of storm-water treatment processes people are using. A hundred grand? Let my kid do that, or I'll do that, but don't spend a hundred grand on it. I'd rather give the money back. So I met with a couple of consultants. I asked if they could do something with floatables. They said they could, and they did a good job, just a pilot study, but it looked promising. So we took that and put together a bigger project on a floatables control project on the canals. And this guy—this guy who called me yesterday—didn't want to get involved. He said it was too much work, so I wrote the proposal. But now he wants his name added, and he wants to make a few changes at the last minute, so forget about leaving at one. I'm stuck there until five. I get a double espresso and that's it; I haven't been down since. I'm just not feeling right. I don't know what it is. Stress I guess."

We follow the increasingly clean water. It moves from the reactors through clarifiers, big, round outdoor tanks. Thickeners are added. The liquid water is siphoned off, chlorinated, and sent down a pipe toward the river. Grit—the sandy stuff that has found its way into the system—goes into big drying piles, then into synthetic bags that are loaded onto trucks and sent to a landfill. The rest of the sludge is piped to the incinerator, close to where we parked. I follow Gordon into the building that houses the incinerator. On a conveyor belt, thickened sludge, looking like mud, rides into the inferno. By far, the worst of today's smells is here, maybe because we are inside, maybe because of the heat. We find a viewing port, three inches across, that looks into the belly of the incinerator itself, where what is left of sewage meets its fate at thirteen hundred degrees Fahrenheit. Here in front of me is a version of hell that Dante missed: burning molten sewage.

"The ash goes to an industrial landfill twelve miles from here," Gordon says. "State law requires that ash be disposed of in an industrial landfill. I don't know why. This stuff is not as bad as industrial ash. It could be used somewhere. This is good stuff." Each day, seventy tons of this good stuff, enough to fill seven or eight trucks, is sent to the landfill. Meanwhile, the land surrounding the sewage treatment plant continues to sink. The cypress trees continue to die. I wonder again what Mike would think of this.

Before leaving, I stop, with a certain sense of poetry, in the men's room. Outside, a fine mist continues to fall from the stacks onto Gordon's car.

"It's about a two-mile run to the river," Gordon says. While we drive, we talk briefly about the SSERP, the Sewer System Evaluation and Rehabilitation Program. This is required by a consent decree with the Environmental Protection Agency, the result of leakage and reporting problems that violated the Clean Water Act. Copies of the consent decree can be purchased from the government for a mere $785. Color maps are available for an extra $325. For New Orleans, the consent decree was somewhat more expensive. The city agreed to a $600,000 fine and a requirement for $600 million worth of repairs. Now, Montgomery Watson, a large engineering firm with a reputation for environmental work, is managing SSERP.

"There's a variety of ways to find leaks in pipes," Gordon says. "You can use smoke. You can use dyes. You've got television inspection. What we did was divide the sewage system into ten sections and evaluate each one. This could be a good thing. We needed to do it."

Most of the fifteen hundred miles of sewage pipes that run beneath New Orleans are my elders. This problem is not unique to New Orleans. Nationwide, six hundred thousand miles of sewer lines grow older every day. In 2000, 8 percent of the lines were believed to be in poor shape. Within a decade or so, fully half of the lines will be in poor shape. "Poor shape," in this case, is a euphemism for leakage. Sewage can leak out, and, just as bad, storm water can leak in, overwhelming treatment plants during periods of heavy rain and leading to undesirable and illegal discharges. The nation will have no choice but to invest in what has been called one of the biggest civil works projects since the federal highway system was built.

We park near the corner of Peters and Delery, near two fifty-four-inch pipelines that discharge water from the sewage treatment plant into the river. A sign warns kids not to play in the area. Kids, playing under the sign, wave when we walk up the levee. The sun shines. A light breeze blows. The river flows past. Water, from the pipes, goes back to the river, having come full circle in its detour through New Orleans. In most respects, the water goes back to the river cleaner than it left the river. One could, in theory, drink the water coming from these pipes. One would not, in theory or fact, want to drink untreated water from the river.

Just downstream, a man fishes from the bank. "He's all right," Gordon says. "It's not like this is an industrial discharge."

We watch a barge headed downstream, and another, and a third pushing its way upstream. The barge headed upstream, making perhaps four knots against the current, will pass Jackson Square, cross under the Mississippi River bridge, ride past Audubon Zoo, and reach the water intakes, where Gordon and I started earlier this morning, in about two hours.

CHAPTER FIVE

THE SHADOW OF MOUNT OII

In the early 1980s, the air stunk around the small mountain of waste. This was not your normal Los Angeles stink, but something more insidious. To the extent that a smell can be sinister, it was. The smell came from the Operating Industries Incorporated Site, more often called simply the OII Site, an abandoned mound of garbage ten miles east of downtown Los Angeles, about midway between the surf beaches and the mountains. In part, the smell came directly from the landfill, slithering down the two-hundred-foot slopes to homes with yards abutting the toes of the mound. But the smell also emanated from the ground around the landfill, outside of the perimeter of the site itself, in some cases sneaking through the foundations of homes, right into the living rooms of middle-class America.

As troubling as the smell was the water. Rainwater falling onto the site seeped through the waste, in some cases reemerging in springs and seeps that trickled down and out into yards, forming puddles next to barbecues and dog runs. As often as not the water found its way into the groundwater, and from there potentially into drinking water. It carried with it a cocktail of contaminants that had accumulated over the landfill's thirty-six-year history, a blend of methane, benzene, vinyl chloride, trichloroethylene, and toluene. Even with water, this is bad stuff. This is the stuff of nervous system and bone marrow problems, liver and lung damage, abnormal heartbeats, dizziness, sleepiness, anemia, various cancers, and, of course, death.

The ground on and around the site was not the sort of ground where most families would want their children to play. Some contend that it was not fit even for animals. In a single cul-de-sac adjacent to the landfill three families faced cancer, and throughout the neighborhood pets developed what residents have called "inexplicable tumors and growths." There were reports of underground fires causing subsidence on the surface and of flames spontaneously shooting up from the ground. In an official release, the federal government pointed out "the potential for an explosion or fire at the site." In 1984, the same year that the site closed as a landfill, the state of California placed it on the California Hazardous Waste Priority List, and in 1986 the federal government's Environmental Protection Agency added the site to its own National Priority List.

Twenty years and something like six hundred million dollars later, the air no longer smells, rainwater no longer carries away contaminants, and reports of fires have ceased. The threat of explosion has disappeared. Plants grow on top of the site. By and large, in twenty years the OII Site metamorphosed from a nightmare to a contained and managed problem, from something close to disaster to something close to a miracle, and in the process became a poster child for Superfund, the commonly used name for the Comprehensive Environmental Response, Compensation, and Liability Act.

To understand the OII Site and its transition from nightmare to poster child, one has to understand the setting. This means, in part, understanding Los Angeles itself. There are, of course, the palm trees and the traffic. Within the city limits, spreading outward from a downtown of skyscrapers, there are close to four million people. If the surrounding area is included, the number jumps to fourteen million. For perspective, this is three times more people than live in all of Oregon and two and a half times more than live in all of Arizona. They could pack the Superdome two hundred times. Most seem to live in a sea of homes with tiled roofs, on yards not much bigger than postage stamps, interspersed with shopping centers, office buildings, highways, industrial facilities, and oil wells. The oil wells are an important part of this landscape. The first one was drilled in 1892, reportedly with a sharpened eucalyptus tree for a drill bit, striking oil only a few hundred feet beneath the surface. By 1897 there were two hundred wells. By 1920 a forest of derricks had sprouted and matured on Signal Hill, just south of Los Angeles. Today, California still produces 12 percent of the nation's oil. Refineries and petrochemical plants spring up in oil provinces, and Los Angeles has been no exception. Currently, over a million barrels of crude oil per day can be refined in Los Angeles.

This many people, undertaking these kinds of activities, produce waste. Enter the OII Site. Before 1948, the OII Site was mined for gravel and sand. What better place for garbage than a gravel-and-sand mine? For four years, beginning in 1948, the OII Site was used for municipal garbage. Starting in 1952, under the ownership of Operating Industries Incorporated, the OII Site accepted industrial liquid waste and sludge. Over three hundred million gallons of liquid industrial waste found their way to the OII Site. All of this was perfectly legal. Waste, after all, had to go somewhere. The site grew to cover 190 acres, equivalent in area to more than 140 football fields. Around it, neighborhoods grew. There are toy stores, grocery stores, chain restaurants, nurseries, a cemetery. Something like twenty-three thousand people live within three miles of the site and more than two thousand people live within a thousand feet. But dismiss all images of an impoverished and downtrodden congregation of worn-out homes and down-on-their-luck residents living nose to nose with an abandoned gravel-and-sand mine full of waste: Homes in the neighborhood closest to the OII Site sell for more than half a million dollars. The view of the landfill comes free with the price of the home.

When I see the OII Site for the first time, I immediately dub it Mount OII. While it pales in comparison to the snow-covered San Gabriels to the north, it rises abruptly above its immediate surroundings, something like the village mounds built by pre-Columbian tribes in the Mississippi River floodplain. To think of it as merely a site—"the OII Site"—discounts its reality. It is not only its height and its steepness that draw the eye, but also its greenness. The Environmental Protection Agency, as part of its cleanup effort, covered the mound of waste with soil and seeds. In an ocean of densely developed land, the houseless expanse of grass and weeds that cover Mount OII has the look of a very steep and very unusual city park.

I drive on streets with names like Liberty Avenue, Rio Blanco, and Bunker Hill. American flags decorate porches. Terra-cotta gnomes and conspicuous signs advertising expensive security systems guard properties. Minivans sit in driveways. Clean white stucco covers walls and brick-red tiles cover roofs. Palm trees and flowers decorate yards. This is a pocket of

I've-made-good America, a place where a family can live, where traffic noise presents nothing more than a minor inconvenience, an idyllic headquarters from which one can commute back and forth to well-paid jobs. Separating all of this from Mount OII are two fences. Signs on the outer fence say, in capital letters, "PRIVATE PROPERTY. CITY OF MONTEBELLO. NO TRESPASSING. NO DUMPING. SECTION 374 B CALIF PENAL CODE STRICTLY ENFORCED." In places, the outer fence comes within a few yards of homes and streets. The inner fence, topped with barbed wire, says "Caution" and "Precaucion," warning adventurers in English and Spanish.

I walk around with a notebook asking residents about Mount OII. I approach a sixteen-year-old in a white sweatshirt with a black knapsack, clearly just returning from school, and ask if he knows that he lives near a Superfund site.

"A what?" he responds.

"A contaminated site," I say.

"No, I didn't," he answers, turning to look up at Mount OII.

A middle-aged man leading a three-year-old girl around his yard knows nothing about the site, although he could easily shot-put a full garbage can from a window at the back of his house and have it land squarely on the toe of Mount OII. A kid playing basketball just shrugs when I tell him the site is contaminated. A peroxide blond dressed for the gym knows nothing about the site. A man cleaning his garage tells me that he thinks the site was once "a dumpster," but that now it has a gas pipeline running through it, carrying gas, he thinks, from wells to the north. He is surprised when I tell him that there are indeed gas lines, but that the gas they carry comes from within Mount OII itself, from a system installed under orders from the Environmental Protection Agency.

Within an hour, I feel like Jay Leno on the *Tonight Show*, looking for Californians who might be able to answer simple questions like "What country borders California to the South?" and "Which ocean washes up on the beaches of Los Angeles?" but getting only blank stares or answers like "Canada" and "A cold one." I wonder if Californians are simply not paying attention. But then another explanation occurs to me. Could it be that the residents know nothing about Mount OII because they do not care, and that they do not care because this place has been, for all practical purposes, fixed? Has Mount OII gone so fully from nightmare to miracle that people living in its shadow have moved on to other things?

Superfund drove the metamorphosis of Mount OII, changing it from a mound of stinking waste leaking toxins onto neighbors' property to an unusually steep hill covered with plants and fortified by fences. But understanding how this happened is like understanding Mount OII itself: First one has to understand the setting in which all of this occurred.

Superfund, or something like it, was inevitable. Industry generated waste, and waste had to go somewhere. The somewhere that jump-started Superfund lies across the country, at Love Canal, in Niagara Falls, New York. Despite its name and its location in Niagara Falls, there was nothing loving or lovely about Love Canal, it had nothing to do with honeymooners, and it was never really a canal. Late in the 1800s, William Love designed a canal to join Lake Ontario and Lake Erie, and in 1894 he started to dig. But like so many projects, it was never finished.

The important realization here is that the site was designed to convey water; it was meant to move fluids, not to contain them.

Like the gravel-and-sand pit that had attracted waste at Mount OII, this uncompleted canal, this hole in the ground, attracted waste. By the 1930s, Love Canal was an industrial waste site. In 1942, it was leased by the Hooker Electrochemical Company, specifically for dumping chemical waste; Hooker Electrochemical Company bought the site outright in 1947, a year after court records showed that muskrat and watercress, once common at Love Canal, were gone. Among the waste that it received was chemical warfare material from the Manhattan Project, as well as pesticides, acids, and solvents. More than four hundred chemicals found their way into Love Canal. The mix caused fires and occasional explosions. Then, in 1953, Love Canal was sold to the Niagara Falls Board of Education. Nothing illegal occurred here. By the standards of the time, nothing irresponsible happened, either. Feasibility studies showed that the canal had been cut into clay soils through which liquids would not readily flow, and only six houses stood anywhere near the site. More clay was brought in to encase the waste. Arguably, Hooker Electrochemical Company had gone beyond the standards of the time, meeting criteria of waste disposal laws that would not be passed for another twenty years. Even the sale of the land to the board of education was above board. The owners went to great lengths to warn the city about the property's history, initially refusing to sell at all because the company thought that building on the site would lead to problems, but ultimately they succumbed to government pressure and transferred the land to the Board of Education for a one-dollar fee.

The Board of Education removed enough contaminated material to fill three Olympic-sized swimming pools before going forward with what was, in hindsight, a monumentally stupid idea: They built an elementary school and sold some of the surrounding land for residential development. In 1957, sewer lines were laid, cutting through the clay soils that had previously trapped the waste within the never-completed canal.

In 1968, road construction created more openings in Love Canal. Children playing in the area developed rashes. By late 1976, after heavy rains, homeowners in the neighborhood that had grown up around Love Canal reported chemicals seeping into their basements. The canal, designed and abandoned eighty years earlier, was finally conveying fluids.

Lois Gibbs, suburban mother of a six-year-old boy and sister to a biologist employed by the State University of New York at Buffalo, lived here. When her son started school, he developed asthma and suffered from occasional convulsions. Acting on information provided by her brother, Gibbs tried, unsuccessfully, to remove her son from the school. In June 1978, she and her neighbors formed the Love Canal Homeowners Committee. She met with the state's commissioner of health and the school was closed. The commissioner referred to the site as a "grave and imminent peril." Even more ominously, the commissioner made two recommendations: "Pregnant women living at 97th and 99th Streets and Colvin Boulevard [should] temporarily move," and "approximately twenty families [should] relocate temporarily any children under two years of age."

Trust in government decisions was not especially high at this point. Identification of eleven suspected or actual carcinogens, including dioxin, accompanied claims of high rates of birth defects, miscarriages, rashes, and cancer. Air pollution levels, affected by vapors rising from the ground, were as much as five thousand times higher than recognized maximum safe levels. Protest signs showed up, like one hanging from a brick home saying, "Evacuate us Now! Federal Aid . . . Please," and another, worn on a young protester's chest, saying, "We've got better things to do than sit around and be contaminated!!" Governor Hugh Carey told residents, "You will not have to make mortgage payments on houses you don't want." The state paid moving and temporary living expenses. Early in the process, 24 residences were thought to have problems; within a year, the number had grown to 237, and within two years to over 1,000.

Congress noticed. Three bills were floated, two in the House and one in the Senate. The first focused on oil pollution, the second on hazardous waste, and the third on all releases of hazardous chemicals, whether the release was waste or not and whether it was oil or not. By the fall of 1980 it was clear that none of the three bills would pass. November rolled around, in an election year, and Congress went into lame duck mode. By this point, recognition that America had a real problem with dangerous waste products had broadened well beyond Love Canal. The Environmental Protection Agency, then only ten years old, determined that more than thirty-nine thousand disposal sites in the United States should be examined. Congressman James Florio, a New Jersey Democrat, pushed a compromise bill that would prove to have broad implications: "In this way," he said, "we can get on immediately with the business of cleaning up the thousands of hazardous waste sites which dot this country." The Comprehensive Environmental Response, Compensation, and Liability Act, often abbreviated as CERCLA, was passed and signed into law by Jimmy Carter. Among the law's provisions was a federal tax on chemical manufacturers, importers, and oil refiners that, coupled with other federal money, created a fund of more than one and a half billion dollars for site cleanup—hence the law's common name, "Superfund."

Superfund required the federal government to identify and take action at sites where hazardous substances might find their way into the environment. But someone had to pay for this, and the one and a half billion dollars earmarked for site cleanup would not come close to covering the cost. So the law also allowed the federal government to pursue what became known as Potentially Responsible Parties, or PRPs. Companies would be held accountable. Superfund called for strict liability, putting Potentially Responsible Parties on the hook even if no laws had been broken. And it called for retroactive liability, putting companies on the hook for actions of the distant past. And, perhaps most frustrating of all to corporate America, Superfund called for "joint and several liability," putting Potentially Responsible Parties on the hook for an entire site even if they had contributed only a small portion of the waste on the site, and, in some cases, even if they had done nothing more than unwittingly buy a piece of contaminated property.

Very quickly, the number of sites specifically singled out by Superfund grew. Within two years, 418 sites had been identified. Within three years the number rose to 551. By 1994, the number rose to 1,300. Superfund quickly became the Environmental Protection Agency's biggest and most expensive program. By 1994, more than eighteen billion dollars was gone and only 12 percent of the sites had been cleaned up. Work was costing, on average, between twenty-five and thirty million dollars per site. Site cleanups typically required twelve years of work. Three cleanups could span the entire career of an engineer, or a bureaucrat, or a lawyer. Looking ahead, the total cleanup bill for all Superfund sites over the next fifty years is expected to be between about a hundred billion and a trillion dollars.

Under the law, most of the money comes from the entities deemed responsible, typically companies that generated, dumped, transported, or sold the material being cleaned up. To think that companies would happily write checks for these cleanups would be more than a little naive. Bob Sablatura, writing for the *Houston Chronicle* in 1995, called Superfund "a bonanza for lawyers," noting that the law had helped just under five billion dollars find its way from corporate accounts to lawyers' accounts between 1981 and 1995 and that additional large sums went to lawyers employed by the Environmental Protection Agency.

Mort Mullins, representing the Chemical Manufacturers Association, after describing Superfund as "one of the largest public works programs in the nation," said that working on Superfund sites was like "trying to use lawyers to build the interstate highway system rather than contractors."

"The reason they hire the lawyers," says Thomas Harrison, a Connecticut-based lawyer, "is that they understand correctly that the liabilities under the law are so expensive; you can't simply sign a check." He points to the joint-and-several-liability clauses of the law. "Even if you have a thimble full of wastes," he says, "it doesn't matter. You are still liable for the costs. It is so draconian; you have no choice but to fight it."

Critics of Superfund abound, and their words are often preserved in court records. In *HRW Systems, Inc., v. Washington Gas Light Company*, the record says, "the legislative history of CERCLA gives more insight into the 'Alice-in-Wonderland'—like nature of the evolution of this particular statute than it does helpful hints on the intent of the legislature." In *Artesian Water Company v. New Castle County*, the record criticizes Superfund "for inartful drafting and numerous ambiguities attributable to its precipitous passage." The record for *United States v. Wade* adds that Superfund's history is "unusually riddled by self-serving and contradictory statements." Most poetic of all, the record for a case involving the Acushnet River and New Bedford Harbor talks of the "difficulty of being left compassless on the trackless wastes of CERCLA."

But there is another possible attitude, and one that might be popular in neighborhoods like, say, Love Canal—or neighborhoods near Valley of Drums in Kentucky, where seventeen thousand drums were unearthed in the early 1980s; or neighborhoods near, say, Mount OII; or among any of the neighborhoods surrounding the ninety-six other Superfund sites in California. Ask the one out of four Americans living within four miles of a Superfund site, and you might find that they are more worried about their own well-being than about corporate complaints and the Alice-in-Wonderland—like nature of a law that will clean up someone else's mess and end its encroachment on their properties and their lives. Karen Florinin, a lawyer for the nonprofit Environmental Defense Fund, says it best. "So what?" she says, responding to corporate complaints about Superfund. "They made a mess; they are being asked to clean it up. Superfund does not send anybody to jail. Superfund does not propose fines or penalties. Just clean up the mess you made. It is not meaningful to ask the question whether it is fair in the abstract. You can only ask if it is more fair than the available alternative. And the answer to that question is yes."

Lance Richman is a geologist, once employed by an oil company but later attracted to an environmental job, seduced by a greener future as a project manager for the Environmental Protection Agency. He is in charge of Mount OII and one other similar site, in Guam. He tells me, soon after I meet him, that he no longer uses his technical talents as much as he once did. "But," he says, "I have access to good talent."

Late in the afternoon, in Lance's rental car, we drive through the gated entrance to Mount OII and head up the slope on a paved road. There are certain safety routines: We wear hard hats and safety glasses, although exactly why we do that here on the face of Mount OII is not clear, and Lance periodically picks up a walkie-talkie to report our location. Presumably, someone in the site's control room is listening, but if so, they do not answer. The road meanders up the face of the mound, which is far too steep for a frontal assault. But even with the meanders, the car's engine strains, and in places we hit dips in the road where car and pavement meet.

On the surface of Mount OII, pipes run everywhere, a plumber's dream, some lying directly on the ground, others raised a foot or so above ground on stanchions. Light brown pipes carry gas, green pipes carry storm water, and small black rubber pipes carry leachate—the liquids pumped from underground, from the intestinal tract of Mount OII. In places, gauges and sampling ports and valves with colored handles stick up from the pipes. We pass bulldozers, portable compressors, and tank trucks. We pass racks of hardware—pipes and spools of yellow cable and spare valves. As we move higher, the sea of red tile roofs that is suburban Los Angeles grows, only to fade out in gray smog.

People whom Lance describes as "Cal Davis eggheads" designed the site. Jorge Zornberg, at the University of Colorado, has said that this is "the first Superfund site with an EPA-approved evapotranspirative cover," referring to the site's reliance on plants to prevent water from mixing with waste.

"This site is not fully lined," Lance says, "which makes it unusual for a hazardous materials site." Most sites like Mount OII are completely lined with tough impermeable plastic sheets covered by soil, in a sense encasing waste in huge garbage bags. "What really makes it work is the plants," Lance says. "It's the wet season, and everything is green up here. Water that falls on the site is wicked up by the plants, so no water moves down into the waste." Anything that could reach through the soil to the waste, allowing rainwater to drain into the waste, would be a problem. Burrows from rodents would cause a problem. So would the roots of some trees. A planted facade keeps the site green, but this is no place for a Mike Rolband wetland. Water is, in this place, the enemy.

As Lance says, this is what passes for wet season in southern California, but the ground is dry. In the concrete gutters along the edges of the road, designed to trap rainwater and move it off of Mount OII before it can soak into the ground, there are no puddles. Importantly, there is no sign of leachate. And there are no odors. And neither explosions nor spontaneous fires belch out of the ground. Dante would be disappointed, but the Environmental Protection Agency is proud.

We reach level ground at the top of Mount OII. "We're on what's referred to as the top deck," Lance tells me. Five feet of soil and a carpet of plants separate us from more than twenty stories of waste. Lance reports in on the walkie-talkie, but again there is no answer. I can hear the traffic from the Pomona Freeway, far below. A softball diamond could be laid out up here on the flat surface of the top deck, but foul tips and home runs might go over the edge, accelerating down the steep slopes and bouncing into traffic on the Pomona Freeway. A better sport might be falconry, which would keep tunnel-digging rodents in check.

We walk to the edge of the top deck and look down on the Pomona Freeway. Plants cover the slope, similar to those that cover the top deck, but on the slope there are patches of bare ground and streaks of eroding earth.

"If you fall from here," Lance says, "you will fall all the way to the freeway." If one had a reason to climb this slope, and had a penchant for trespassing, one would want to bring ropes.

"People are concerned about what will happen in an earthquake," Lance tells me. "Will the slope collapse and slide onto the freeway? We've seen some cracking on the slope, but only a little more than we originally expected. The sides of the landfill are different than the top. It's a different design. Every five feet we have a tough plastic liner built into the slope. Tiebacks give it stability. It is similar to terraces, but it would not be right to think of it as terraces, exactly."

The Pomona Freeway is separated from Mount OII by a fence, barely visible from here. Lance tells me that the fence, heavily reinforced, could intercept material sliding off the slope, breaking the fall of a landslide before it reaches the freeway. "This is by definition a mobile system," he says. "Nothing quite stays in place." Lance gestures with his hands and sways his torso, looking like a sailor on the deck of a storm-tossed ship, underscoring the site's instability. "I can show you pipes," he says, "with paint lines above the surface, where the ground has subsided around the pipe."

In effect, the garbage buried here settles and decays, and the ground on top subsides. In places, Mount OII can lose one foot of elevation in a year. Slowly, the mountain shrinks.

Back in the car, we drive across the top deck and down to stop at what Lance calls an air dyke. "On most of the site, we apply suction to the ground," Lance says, "bringing up gas and liquid and piping it down to the treatment facility. The air dyke here on the edge of the site pumps air into the ground, making a wall of positive pressure. It's injected about fifty feet down. Just enough pressure, but not too much." He points out a pressure gauge that reads eighty pounds per square inch. An underground breeze drifts through the innards of Mount OII; air pumped into the ground here blows through the buried waste toward the suction, becoming increasingly vile before being slurped into pipes and sent to the treatment facility down below. In a less than subtle twist of Los Angeles irony, just to the other side of the air dyke, just off the edge of Mount OII, an industrial facility pumps natural gas in and out of the ground, where it is stored in underground reservoirs, thousands of feet below Mount OII, before powering electrical generators that feed the lights and refrigerators and air conditioners of Los Angeles.

We look at a nest of three vertical pipes, standing at attention. Lance calls them "points of compliance wells." They reach downward into the waste, each tapping a different depth. "There are thirty or thirty-five scattered around the site," Lance says. "We can monitor pressure, gas, temperature. Depending on what we see, readings from these wells can trigger actions. The readings help us with decisions about what to do, about how to spend money."

A hawk hovers above us, as though it has spotted some prey, and Lance mentions that the hawks sometimes land on valve handles. "They're big enough and heavy enough to turn the smaller valves," he says. From behind the hawk, I can hear the noise of an airplane, and south of us a helicopter flies past. Gas hisses as it moves through a pipeline.

We climb into the rental car again but stop once more to look down on the neighbors as they water plants and play basketball and commute to and from work or school, unaware that they are watched. "Early on, when we first became involved with the OII Site, we did an emergency response," Lance says, pointing to the bottom of Mount OII, into the tiny backyards that form a thin margin around the red-tiled roofs of the nearest homes. "We built a toe buttress to keep the landfill from falling into people's homes." I strain to see the buttress separating

Mount OII from its neighbors, but from here it is no more visible than the air dyke.

The Environmental Protection Agency is the public face of the OII cleanup. Although they oversee operations, they do not run the project. Under Superfund, this is the norm rather than the exception. The Environmental Protection Agency identifies Superfund sites, then identifies anyone involved with the hazardous materials causing the problem. They reach out to Potentially Responsible Parties, the PRPs.

This is not without pain. Potentially Responsible Parties do not necessarily want to be involved. The problem in question, as often as not, was generated by past practices undertaken without breaking any laws. Employees who might have been involved with the problem have moved on or retired. Thomas F. Harrison, the same attorney who said that liabilities under Superfund are so expensive that companies cannot simply sign a check, also believes that few Potentially Responsible Parties intentionally broke any laws. "Most of the time," he believes, "the polluter is not what people usually think of—someone dropping off a drum in the middle of the night. It is some business doing what they believe to be the perfect lawful thing." But now Superfund, for all of its ambiguities and for what to the Potentially Responsible Parties may appear to be a total absence of fair play, is the law. While the exact level of responsibility and cost may be an issue, the only real way out is to show that there was no involvement. If there was involvement, a Potentially Responsible Party becomes a Responsible Party, an RP, and gains the privilege of participating in the cleanup.

Potentially Responsible Parties and Responsible Parties can be companies, but they can be individuals and government entities, too. Almost anyone identified as a Potentially Responsible Party or a Responsible Party, for having been involved in what one thought was a perfectly acceptable practice, is going to be upset. Some feel strongly. Example: Following an Environmental Protection Agency decision about the Beede waste-oil dump in New Hampshire, more than a hundred owners of small service stations, accompanied by their families, rallied in front of Boston's JFK Federal Building, chanting, "No way. We won't pay." Usually, though, major Responsible Parties are large companies, often with names recognized across the nation and around the world—big chemical companies, big oil companies, public utilities, trucking companies, airlines. As early as 1994, a survey showed that 367 of the Fortune 500 industrial and service companies had been identified as Potentially Responsible Parties. It is reasonable to bet that any large company that deals with chemicals is a Responsible Party at one or more sites somewhere in the United States. And it is safe to bet that they do not receive many sympathy cards when they complain about Superfund or many congratulatory notes when it comes to their efforts to clean up sites. From a corporate perspective, legal compliance and radio silence may be the best strategy when it comes to Superfund involvement. And so the Environmental Protection Agency is the public face of Superfund, but the Responsible Parties can run the cleanup itself.

New Cure, Inc., a company established by the Responsible Parties solely for the purpose of cleaning up, runs Mount OII. New Cure doesn't talk to journalists. In this regard, the New Cure representative, Les LaFountaine, is polite and even friendly, but firm. And who can blame him? What have the Responsible Parties got to gain? Even if a journalist writes only good things about the cleanup effort, there is still the stigma of involvement with hazardous chemicals, and no company wants to rub the noses of its shareholders in the expensive liabilities of contamination. So New Cure, Inc., and companies like it quietly represent Responsible Parties and clean up the messes while happily letting the Environmental Protection Agency act as the public face. Aside from the minor detail that no one really wants to be in this situation to begin with, everyone wins.

For Mount OII, as a matter of curiosity, I track down names of the Responsible Parties, understanding from the outset that identification as a Responsible Party does not imply that anything untoward occurred. These organizations could be Responsible Parties even if they had consistently behaved both legally and responsibly. While it is easy to find things like site design specifications and site histories, finding a comprehensive list of Responsible Parties is not so easy. But I find a few names. Chevron is one. The navy is another. The California Department of Transportation is another. Air Products and Chemicals, Inc., is another. IT Corporation is another. The list includes Unocal, Exxon, ARCO, and Texaco. There are over one hundred names. Records indicate that nearly four thousand companies contributed hazardous material to the site, ranging from a small quantity up to 15 million gallons. Many of the companies are no longer in business. Some—those who contributed less than 110,000 gallons—were identified as what the Environmental Protection Agency calls "de minimis" parties and were dismissed after paying a small settlement. The larger contributors are in for the long haul and the full cost, whatever it may turn out to be, of cleaning up.

The tangled web of involvement with Mount OII does not end even here. There are, in addition to the Responsible Parties, organizations that become involved with the cleanup as contractors. There is, for example, the U.S. Army Corps of Engineers, which works with the Environmental Protection Agency to manage contractors and oversee work. There is Foster Wheeler Environmental Corporation, based in New Jersey, that won a forty-million-dollar contract for design and construction of the Mount OII cap, the gas collection system, and the facility that treats the gas from the site. There is, or was, a company called Greenfield Development, once based in Seattle, that had hoped to develop part of the site, after cleanup, as a shopping center, a venture that some past employees claim led to Greenfield Development's demise. And there are nurseries that supply plants, universities that conduct research, laboratories that analyze samples, printers that publish notices and reports, suppliers that provide pipes and paint and heavy equipment. And, of course, there are lawyers.

One cannot spend the six hundred million dollars required for something like Mount OII without contributing to the economy here and there.

The drivers in Monterey Park seem to be among California's worst. On my way to Bruggemeyer Memorial Library, a few miles from Mount OII, I realize that turn signals are considered optional. Stopping in the middle of the road to wait for a parking space is an accepted practice.

The library houses information about Mount OII. A bulbous librarian, gray hair brushed straight back and shedding onto a purple sweater, says that she knows nothing about the site, even after I show her a notice, issued by the Environmental Protection Agency, that guides the public here. Eventually, she phones another librarian, who not only has heard of Mount OII but knows exactly where the information is stacked. "But," she says, unprompted, "we don't have any information about the companies and the chemicals they dumped. These are companies that went out of business twenty years ago. We just don't have lists of the poisons they dumped." While she walks me through two rooms full of books, she tells me that lots of people come around to sift through the Mount OII repository.

Bracketed between books called *Censorship* and *100 Banned Books*, on the one hand, and *The Encyclopedia of Crime*, on the other, over fifty volumes describe Mount OII. Some individual volumes are four inches thick. There is a ring binder holding the *Draft Residential Air Monitoring Plan—Operating Industries, Inc.*, and another holding *Landfill Response to Seismic Events*. There are numerous feasibility studies and Final Records of Decision. There is the *Draft Residential Gas Control Program Plan*, which, over the course of hundreds of pages, recommends sealing "indoor entry points"; that is, it recommends filling in those pesky foundation cracks, because in addition to letting in bugs and moisture they may let in methane and vinyl chloride. There are tables of mind-numbing numbers and photocopied pages of handwritten notes: "Fred indicated that any new emission source such as a leachate treatment facility would have to undergo a new source review process." There is a health study comparing information from 1,349 adults and 472 children living near Mount OII to 928 adults and 434 children living ten miles to the east, noting that "no significant excesses from the site were seen for any major medical problems, including overall mortality, cancer, liver disease, and adverse pregnancy outcomes." But it goes on to say that "two to four times as many people living downwind of the site complained of symptoms as did those who lived furthest from the site or in the comparison area." These symptoms included headaches, eye and throat irritation, nausea, trouble sleeping, and feeling tired. It says nothing about driving habits.

Most of the information here is old news. The reports are about conditions before the Environmental Protection Agency worked its magic, or they describe, in terms of what will be done, what by now is completed. The cap is in place, indoor entry points have been sealed, the leachate treatment facility has been built, and the children surveyed in the health study have grown up. No one I spoke to downwind of Mount OII complained of headaches or nausea or trouble sleeping. But none of this is to say that the Environmental Protection Agency is ready to walk away from Mount OII. As a rule, it is difficult to close out Superfund sites, to say that they are done and to turn them over to local governments that might want to build, say, an office park or a playground or a school. The Environmental Protection Agency claims that many sites will require thirty years of operations and maintenance before they can be considered cleaned up. Understanding just when a site is finished can be confusing, in that there are different steps involved. One step places sites on the "construction completed list," making it sound, to some, as though the job is done. But the goal is to reach what the Environmental Protection Agency calls "deletion." The government deletes sites when the cleanup goals have been met, which may be long after construction has been completed. And even deletion has a catch: If new problems become apparent, sites can be undeleted, and Responsible Parties are back on the hook.

In 2002, almost nine hundred sites were on the construction completed list and fewer than three hundred were on the deleted list. Mount OII did not make either list.

When I ask Lance Goodman how long it would be before Mount OII is finished, his response is simple: "A long, long time."

The treatment facility at the base of Mount OII, accessed through a gate in a chain-link fence, looks like a compact industrial facility. Pipes join tanks and stacks and tubs the size and shape of buildings. Its noise competes with and drowns out traffic. Whoever designed it made a decision to pack things in, so the pieces fit together like a jigsaw puzzle, with just enough walking-around space for access. But this is no industrial facility, at least not in any normal sense. This is the end of the pipe. This is where all of the nasties descending down Mount OII end up. This is where New Cure, Inc., working under orders and guidance from the Environmental Protection Agency, cleans up the stuff thrown away by four thousand companies over twenty years ago.

What appear to be fumes leap from the top of two gray stacks, each as big around as a car. But I smell nothing.

"This is the Thermal Destruction Facility," Lance tells me, raising his voice over the noise. "The gas we put through there is destroyed. It comes out at fifteen to eighteen hundred degrees Fahrenheit." Unprompted, Lance reminds me, several times, that this is a Thermal Destruction Facility, not an incinerator. It does not merely burn waste; it destroys waste. The heat in these stacks would melt silver. A few hundred degrees of additional heat would boil lead. One could roast marshmallows from ten yards away. "It was built," Lance says, "to give us four nines. It is 99.99 percent clean. If you could stand the heat, you would be better off breathing the fumes up there than the Los Angeles air."

Most of what they are burning—rather, most of what they are thermally destroying—bubbles up out of the leachate piped down from Mount OII. The leachate, the intestinal bile from the guts of Mount OII, flows into what Lance calls a spurge tank, a structure that looks like a cross between a bunker and a

Roman bath, with a sign painted on its side saying, "Sequencing Batch Reactor Well T-8A." This is where the liquid gives up its gas. Before injection into the Thermal Destruction Facility, the gas is dried, so no steam billows from the top of the stacks.

The leachate that remains behind, it turns out, is not as nasty as one might suspect. "We call it a 'poor sewage,'" Lance says. "There's not enough meat to eat for bacterial treatment." Water treatment, like that used at the East Side Treatment Plant in New Orleans, relies on bacterial degradation, but with the bile from Mount OII, degassed, sugar has sometimes been added to promote bacterial growth. Leaving here, the liquid leachate, cleansed and tested, goes to the municipal water treatment system. The bile from Mount OII, after a decades-long residence in the guts of the mountain, after a roller-coaster ride through pipes and a brief stay in Sequencing Batch Reactor Well T-8A, after a thorough belching off of noxious gas, joins the household waste coming from under those brick-red-tiled roofs that spread across Los Angeles. The cleansed leachate mixes with dirty bath water, drain cleaner, laundry detergent, dish soap, and the general down-the-drain filth of daily living, all headed together for further cleaning at the municipal water treatment facility.

Lance, in telling me all of this, also mentions that the facility has been overbuilt. It was designed and built to treat what was believed to be in the landfill, but the belches, burps, and bile trapped in the pipes and sent down to the facility are not as bad as what was expected. The leachate is so clean that it will not support bacteria and the fumes coming from the stacks freshen the air of Los Angeles.

It occurs to me that the Responsible Parties, who, human nature being what it is, may have been reluctant partners in the first place, might be frustrated with their investment in the site and the requirements established by the Environmental Protection Agency. Could this be overkill? Did Superfund's demands go too far?

"Overkill?" he says. "You would want to talk to the public about that. The public thinks differently about these things than the Responsible Parties." He pauses for a moment, reverting back to low-key Californian, becoming pensive. "This," he says, "was the right thing to do."

A hawk circles well overhead, watchful, perhaps taking advantage of upwellings from the Thermal Destruction Facility stacks or perhaps reveling in its fresh air. If it screeches, I cannot hear it above the noise that surrounds the plant.

We walk around Sequencing Batch Reactor Well T-8A and enter a combined control room and laboratory. Switches and gauges and graphical printouts cover panels in one room. There are, too, the official-looking red buttons that one is always tempted but never quite willing to push. Beakers and flasks dry in another room, next to automated equipment used to test water samples. Not long ago, when Mount OII was a teenager, well-equipped research laboratories could not have detected the levels of contamination that can be measured by this equipment, a fact that bespeaks the world's newfound fascination with contaminants.

A tinny version of Aerosmith coming from a clock radio permeates the room, and the one man who works here, a New Cure employee, nods hello but otherwise ignores us. Lance says that a government employee visits several times a week, to keep an eye on things. From what I can see, the visits would quickly become a boring chore. After seeing all of the pipes and the cap and the air dykes, after looking down from the top deck, after seeing the heat rising from the Thermal Destruction

Facility, the control room and laboratory are best described as anticlimactic.

Outside, Lance has one more thing to show me. "We did something here we are proud of," he says. "We have six microturbines. We cherry pick gas off the landfill and generate electricity for use on site. We could sell it to Pacific Gas and Electric if we didn't need it all here." A gauge on one of the units tells me that it is running at 67.3 kilowatts. Combined, all six microturbines could provide enough electricity to run forty thousand bright light bulbs. With this power, one could light up a small town. Alternatively, one could pump air into the ground and suck bile from the core of Mount OII.

Standing in the shadow of Mount OII, speaking more generally, Lance states the obvious. "This is a first-class operation," he says. "Whenever Senator Boxer goes on the Hill, this is the site she talks about for Superfund success. If you can imagine, this mountain was just a two-hundred-foot-tall pile of garbage when we started, leaking contaminants."

I point out that anyone buying a home next to a pile of garbage, even twenty years ago, might have expected problems. For the second time, Lance seems momentarily incredulous. "I would argue," he says, "that homeowners didn't know what they were getting into. They didn't know it contained hazardous materials."

Near the end of our conversation, he reiterates his pride in the site. "The people around here," he says, "are just effusive about this site."

Effusive? The people who live in the shadow of Mount OII are better than effusive. They are ignorant. They have moved on in life and are worried about other things. The landfill that leaked into their backyards two decades ago has become an odd-looking mountain, just part of the scenery in urban southern California, blending in with the highways and cars and red-tiled roofs and rocking oil pumps. New Cure, with the Environmental Protection Agency looking over its shoulder, keeps the mountain in check, spending the Responsible Parties' money to contain and pump out and clean up the bile and the belched gases. Perhaps the Responsible Parties occasionally complain about the unfairness of a law that forces them to pay for something that happened a long time ago and that, at the time, was a perfectly acceptable practice. But, if so, their complaints are mumbled in private or reserved for a trial judge. For the most part, they focus energy that could be used for complaining on running a first-class operation, resigned to the reality that cleaning up is not, after all, an issue of fairness in the abstract, but rather just a fact of life, a part of doing business.

It is late in the day when I drive away from Mount OII. The sun is setting, turning the smog from gray to pink. Lights come on in homes. Above the traffic, above the silhouette of Mount OII, timid pinpricks of light become bold in the darkening sky. Smog and the headlights of a million cars are not enough to blank out the brightest of the stars.

CHAPTER SIX

WASTE NOT, PROFIT NOT

I follow Sean Skaling into an office building in Anchorage, Alaska. His task here is to inspect the building, to assess whether or not it measures up in terms of environmental performance. Sean, in his thirties, is boyishly handsome, with thick dark hair and, almost always, with the hint of a smile shining at the corners of his eyes. He runs an organization called Green Star, a nonprofit, like Greenpeace and the Natural Resources Defense Council and the World Wide Fund for Nature. Unlike these mega-nonprofits—groups sometimes called Big Green, in reference to their environmental aspirations, their size, and their bank books—Green Star is small, consisting of Sean and two others, one of whom works part-time. Green Star's total cash flow would be less than decimal point noise in the big budget world of Big Green. And while Green Star is based in Alaska, it leaves protection of polar bears and wolves and whales and unspoiled wilderness to others. It abstains from in-your-face environmentalism. Instead, Green Star helps retail stores, restaurants, service providers, oil companies, and nonprofits understand how to improve their environmental performance, mainly by managing their waste streams. That is, they help manage what was once known as and thought of as garbage. What Green Star lacks in financial clout, it does not make up for in glamour.

"Green Star is not prodevelopment," Sean tells me, "but we're not antidevelopment either." He laughs. "We're for smart development, with the emphasis on smart, not development. We stay apolitical."

The Green Star commandments might read like this: Waste not. Avoid using what you do not need. Reuse anything that can be reused. If something cannot be reused, recycle it. Throw it away only as a somewhat shameful last resort.

I follow Sean through the building while he checks thermostats and admires low-energy fluorescent bulbs. He has a portable light meter. In a meeting room, he notes that light intensity is 649 lux—light enough, but not too light. He notes, too, that some of the rooms have motion detectors set to turn off lights if the room is empty or its occupants too sedentary. He pops the back off of a wall-mounted emergency light, remarks that the batteries are not rechargeable, and replaces the lid. He checks a copy machine to see if it defaults to double-sided copies. He looks into a recycle bin next to the copy machine. The bin is empty. "I hate to see empty recycle bins," he says.

Later, down the hall, next to another copy machine, he is rewarded: Used double-sided paper fills two bins. In the lunch room, he is rewarded again, this time by washable mugs and plates and utensils. There are no paper cups or plastic forks. Countering this, the absence of a dishwasher disappoints him. "Dishwashers," he tells me, "are much more efficient. They use less water than hand washing." But another reward balances this disappointment. Next to the refrigerator, someone has pinned up a Green Star poster: "Our policy is to improve environmental quality through wise business decisions."

The approach is simple. Businesses pay a small application fee and Green Star assesses their behavior. Green Star provides advice.

"We go to a company," Sean says, "and we look for ways to save them money. Cardboard is a big one for savings. It's bulky and it fills dumpsters. Another one is glass. We have a lot of glass. Glass recycling can pay for itself. We say, 'Hey, you are paying an awful lot to get rid of your garbage. Instead of paying for throwing it away why don't you recycle it? You'll end up paying less and doing the right thing.' That is the easy sell."

In addition to application fees, Green Star gets corporate sponsorships, foundation donations, and government grants. And in addition to solid garbage, the stuff you can hold in your hands and throw into a recycle bin, they are interested in vaporous garbage, the stuff thrown into the air. "The government grants we receive are really contracts," Sean says. "I just got the contract for an air quality project. We want to reduce carbon monoxide around town. Anchorage has problems with carbon monoxide, mostly from cold-starting cars in the winter. About 75 percent of the carbon monoxide is from cars. The goal is to avoid cold starts by getting people to ride the bus. If they're going to drive, they should plug in an engine block heater. We're giving people timers so that they can use block heaters more efficiently—you really only need to run your engine block heater a few hours before you start your car, so that means the timers need to be in weatherproof boxes that mount on the outside of your house. And I think the state or the city has a block heater installation program where you can get an engine heater installed for twenty-five dollars."

Green Star's efforts extend beyond the workplace to individuals, but the same philosophy applies: People can and should improve environmental performance without impacting their bank books. With government support to pay for engine block heaters and free timers, individuals pollute less at no cost. As a bonus, avoiding cold starts extends the life of the car's engine, which saves money and keeps the car out of a landfill that much longer. Costs are not internalized, but pollution is reduced.

Sean talks, too, about two companies that pay employees to ride bikes to work—three dollars a day, right off their bottom line, to get employees on bicycles. He talks about expanding this idea to other companies. He has acquired a grant that will allow him to help companies pay their employees for biking, pulling the cost out of their bottom line. The companies sign up for the Green Star Air Quality Award to become eligible. "It's a great concept," he says. "It's a great idea, and we're curious to see if it is going to take off. So what we are going to do is our first twelve applicants enrolling this year will be eligible to get into a trial. Half of them will pay employees on their own and we will pay the other half with some of our grant money. We want to see if companies will do it without government money."

This is not intended to be a permanent payout. The idea is to change habits and then remove the incentive payment. Get people riding, then cut off the three-dollar reward and see if they stick to their bikes. In the summer—long days with mountains framed against blue skies—three dollars a day is generous. But this is Alaska. People here drive big trucks. They work for oil companies. Winter is dark and cold. Old habits might look comfortable. Three dollars might not be enough.

Outside of the building, we climb into Sean's car. Though it is summer, he himself did not earn the three dollars today.

We circle behind the building and park again. Sean gets out, checks the dumpster, and jots down a few notes.

Green Star operates out of rented space on total annual revenues of about three hundred thousand dollars a year and assets worth about seventy thousand dollars. In contrast, the Nature Conservancy, the biggest of Big Green, has annual revenues of more than seven hundred million dollars, assets worth three billion, something like three thousand employees working from five hundred offices in thirty countries, and an eight-story headquarters building just outside of Washington, D.C. The Nature Conservancy is in what some call the "bucks for acres" business: They buy land for conservation. It has described itself as "radically neutral"—like Green Star, the Nature Conservancy wants to protect the environment but remain apolitical.

From a Nature Conservancy advertisement: "Since 1951, we've been working with communities, businesses, and people like you to protect nearly 117 million acres around the world." For context, Germany covers 88 million acres and Iraq covers 108 million acres. France, at 135 million acres, only marginally outweighs the Nature Conservancy. For now, Texas, at about 170 million acres, claims more land than the Nature Conservancy, but the Nature Conservancy grows every year.

If protecting land is the objective, radical neutrality appears effective. But in the world of Big Green, approaches other than radical neutrality play a role. There is, for example, the Center for Biological Diversity, an organization known for threatening and filing suits against the federal government to force the addition of new species onto the Endangered Species list. It runs advertisements encouraging people to become what it calls "biodiversity activists." It offers specific suggestions: "Tell Bush to provide full protection to orcas" and "Tell California state legislators to protect condors." For encouragement, it celebrates victories with Web-site headlines: "Habitat protection ordered for mountain yellow-legged frog," "Off-road vehicles booted from 572,000 acres of critical habitat," and "Judge okays lawsuit against killer wind turbines." With this last headline, the Center for Biological Diversity provides more detail: "Wind turbines at Altamont have killed an estimated 880 to 1,330 golden eagles, hawks, owls and other protected raptors each year for the past 20 years." Even wind has its problems.

The Center for Biological Diversity is not alone in using the legal system to push environmental progress. Earthjustice, formerly known as the Sierra Club Legal Defense Fund, spends something like eighteen million dollars a year and employs fifty attorneys in nine offices to, in its own words, "reduce air and water pollution, to safeguard national forests, parks and wilderness areas and seashores, to contain toxic materials, to achieve environmental justice and to preserve wildlife habitat." The like-minded Environmental Defense spends another roughly thirty million dollars a year; its literature states, "Since 1967, we have linked science, economics and law to innovative, equitable, and cost-effective solutions to society's most urgent environmental problems." Yet another organization, the Natural Resources Defense Council, spends forty million dollars a year, again using the legal system to improve environmental performance.

With all this money floating around, it is clearly not negative cash flow that makes these groups nonprofits. It is, instead, a tax status. Nonprofits are known to the Internal Revenue Service as 501 (c) (3) organizations. As often as not, they are incorporated, making them 501 (c) (3) corporations. As such, nonprofits do not pay taxes; instead, as an annual obligation to the Internal

Revenue Service, they submit a Form 999, accounting for the money they have raised and spent. Nonprofits make money. To function, they have to make money. But they cannot distribute their wealth for the personal gain of founders or supporters, which means, in the end, that more money remains available to support whatever mission a nonprofit may pursue, whether it is to help corporations manage waste streams, to buy land for conservation, or, through the legal system, to encourage the government to obey its own laws.

Where does the money come from? Mostly, it comes from donors—both individuals and foundations. For something like twenty-five dollars, anyone can become a supporter of Big Green; in return, donors feel like they are making the world a better place, and as often as not they receive a glossy magazine with pictures of endangered birds and giant sequoias and, occasionally, a cheap knapsack with the organization's logo.

Money comes, too, from foundations, from government, and from corporate grants. Something like forty thousand foundations control assets worth nearly a quarter trillion dollars. As registered charitable foundations, they are required to give away more than 5 percent of their wealth each year. Government, too, is generous. The government, according to the *Sacramento Bee*, gives just under four hundred thousand tax dollars a day to about twenty major environmental nonprofits. Ironically, some of this government money goes to those groups that are most critical of government policies. As an example, the Lands Council, a group fundamentally opposed to government policies that allow the sale of timber from public lands, received about a tenth of its budget from the government in 2004; interestingly, the agency providing the grant was the very same agency responsible for selling timber from public lands, the U.S. National Forest Service.

Another bit of irony: On the same day that the National Wildlife Federation sued the Environmental Protection Agency over water quality issues, they applied for an Environmental Protection Agency grant. Eventually, after some time had elapsed for processing grant applications, they got the money.

Where the private sector is involved, irony folds into conflicts of interest. BirdLife International, a nonprofit dedicated to protecting birds and their habitat, got into bed with Rio Tinto, a mining company dedicated to, as one might expect, mining. "It's how the money is used, how it's targeted, and how it's delivered on the ground," claims Jonathan Stacey, a project manager with BirdLife International. "As long as it stays with BirdLife's key objectives, there is a strong foundation for cooperation." Another representative from BirdLife International puts it this way: "There seems to be a misconception. We are talking about partnerships and not sponsorships." They know they cannot stop mining in every case, so why not help the miners manage their operations to minimize impacts and, in exchange, help themselves to some of the miner's money?

From an article in *Ethical Corporate Magazine*: "Yesterday's adversaries are increasingly becoming allies."

From an *Inter Press News Agency* article: "To skeptics, the partnership between multinational corporations and environmental groups is the business of selling 'feel-good conservation' to prop up a company's sagging public image."

Eric Williams, of the public relations firm Environics, describes the more adversarial nonprofits: "They're what we call a 'conflict' industry. They may be nonprofit, but their bottom line is to make money, and they do that on conflicts."

In an E-Wire report, Mike Hardiman, of a group called Public Interest Watch, renders an opinion about a nonprofit well known for applying the principles of civil disobedience to the environmental movement; tactics like trespassing, lying down in front of bulldozers, and using zodiacs to board vessels all become weapons in the fight for the environment. Hardiman's opinion: "Greenpeace has devised a system for diverting tax-exempt funds and using them for nonexempt—and oftentimes illegal—purposes. It's a form of money laundering, plain and simple."

The Economist wonders about accountability in the nonprofit sector: "In the west, governments and the despised agencies are, in the end, accountable to voters. Who holds the activists accountable?"

No one leaves the room unscathed. Deborah Dasch, writing about people working for Big Green, dispels a myth: "Instead of being outdoors hugging trees, these careerists are in the offices of federal and state lawmakers and corporate America, pressing the flesh of the decision makers."

Even environmental nonprofits criticize environmental nonprofits. A young woman from the Nature Conservancy describes what sounds like a personal awakening: "I was in a room with other nonprofits. These were more adversarial nonprofits. I felt like a pariah. I felt like they thought I was evil."

Despite the rain, Green Star's annual Electronics Recycling Event is anything but rained out. In Anchorage's spring drizzle, a fifty-year-old woman pushes a shopping cart full of computer parts, unboxed, across the parking lot. On the curb, a couple, teenage child in tow, wrestles two computers out of the backseat of a car, the job made more complex by the tangle of still-connected wires that join the components to one another, creating a serpentine mess of unwanted circuitry and plastic. Leaning against a wall just out of the rain, a young man with a blue streak in his hair and a pierced eyebrow talks on a cell phone; he stands next to a four-foot-long orange cart, loaded with black-and-white televisions, microwave ovens, and stereos, the average age of which exceeds his own. Inside, in the space left by a now vacant grocery store, a line of people and their carts full of old electronics doubles back on itself. The people lean on their carts, look at their watches, and scold their kids. A shorter line holds more recent arrivals who still need carts. Volunteers in green vests work the crowd, offering instructions and pleas for patience. Other volunteers staff folding tables at the front end of the line, weighing the contents of carts and transferring those contents onto pallets. Still other volunteers shrink-wrap the pallets. And others take money from the people pushing the carts: five dollars to get rid of a television or a computer monitor, five cents per pound for everything else. Businesses pay more: fifteen dollars for televisions and monitors, thirty cents a pound for everything else.

One of the volunteers tells me that her employer initially saw the recycling event as a way to get rid of a few old computers—computers that had become so obsolete that they were no longer suitable even for interns. Her boss had told her, "Just throw them in a van and haul them down." There were too many for a van. Her employer, in the end, rented a tractor trailer.

"The recycling event," she says, "has been more successful than anyone hoped—maybe a little too successful."

The Anchorage library system gave up sixty monitors. The school district gave up five hundred monitors, which, with other components, filled 115 pallets. People have driven over a hundred miles to dump this stuff. The room around us is strewn with wooden pallets covered in the detritus of the electronic revolution—keyboards, televisions, photocopiers, central processing units from main frames, fax machines, projectors, VCRs, hundreds of joy sticks, broken video games. Without prejudice, Green Star volunteers sort it, palletize it, and shrink-wrap it. This junk is ready to go.

To salvage these things makes sense. Laws prevent businesses from sending certain components to landfills, including monitors, which are classified as hazardous waste. But aside from this, these things contain valuable raw materials. Making new computers and the components that go into them cannot be done without environmental sacrifices. Silicon and gallium arsenide are key components of semiconductors, with crystals shaped by hydrofluoric acid. Copper or aluminum, each scraped from the earth and leaving behind their own special scars, makes printed circuit boards; emissions of hydrogen fluoride, arsine, phosphine, and arsenic come from the process; large quantities of water—as much as thirty-three thousand liters per computer—are needed. According to environmental guru Paul Hawken, a ten-pound laptop computer requires the handling and processing of forty thousand pounds of material, much of it in mines cut into the surface of pristine ecosystems in the developing world. Something like sixty million computers become obsolete each year, of which maybe three million are reused or recycled, with the rest thrown out or—for owners who know better than to throw away appliances that contain hazardous waste—in perpetual storage, collecting dust on a back shelf behind the out-of-date company stationary and the five-year-old tax records.

From here, most of this stuff will be shipped south, by truck, barge, and train. Some of it will be disassembled, with the parts being reused and recycled. Some will be recycled without disassembly. Some will be thrown away in special landfills built to accept hazardous materials.

Somewhere behind the scenes, surrounded by all this junk, Sean Skaling is working, too busy to talk to me. Later, after the event, I track him down.

"I did a little of everything," he tells me. "Obviously, we in the office organize the whole thing and try to get all the pieces to come together. During the event, my domain was at the back dock, overseeing the loading of pallets on the trucks and labeling of pallets and calling the trucking company to pick up the trailers. This was our first electronics recycling event ever, and we didn't have any shipping expertise at all. We had the building space donated to us, with six shipping docks. And I had everything scheduled and trailers reserved. Not enough. We ended up needing more. We had eight trailers lined up but needed fifteen. We had everything lined up, trucks scheduled to come and go, these six shipping docks. But what I didn't realize was that it was an older building. I don't know if the trucking standards had changed or the ground had subsided or what, but the trucks just didn't match up to the docks. In some cases it was fifteen inches off. You can't quite drive an indoor forklift over a gap that big. So we had to get these ramps at the last minute. Our goal was to work Friday and Saturday, then take Sunday off and tidy up on Monday. We wound up staying right through Wednesday."

Working castaways would find new homes at charitable foundations. From the no longer functioning machines, copper and aluminum wire would be stripped out and melted down. Glass screens would become grit needed for the lead smelting process. Plastic cases would be regranulated, pelletized, and reused in injection molds or to make park benches. At the end of the two-day recycling event, one hundred volunteers had loaded 280,000 pounds of electronics, shrunk-wrapped onto 554 pallets, into fifteen trucks. Once processed, less than 10 percent of this stuff would wind up in landfills.

In Alaska, Green Star is one of more than fifty active environmental nonprofits. There is, for example, the Alaska Center for the Environment, formed in 1971, known for having sponsored the Kenai Brown Bear Festival in 1998 and for managing a popular summer camp. A Web site provides advice for their nine thousand members: "Let your Assembly members know that you DO NOT SUPPORT the proposed changes to the pesticide ordinance because you do not want to weaken our right-to-know about what pesticides are being used around us." This advice comes with a sample letter; all a member does is fill in the blanks, print the letter, and send it to the appropriate representative, whose name and address are, of course, provided. The Web site offers advice, too, on protecting roadless wilderness: "Please contact local land manager Joe Meade, Forest Supervisor on the Chugach National Forest at 743–9500 and express your support for protecting Roadless areas within Alaska's favorite forest! Tell him that you love the Chugach and to keep it Roadless!"

The Alaska Conservation Foundation, another of Alaska's nonprofits, has a Web site that includes a convenient "donate now" hot button. The Alaska Conservation Foundation is, according to the American Institute of Philanthropy, one of the nation's best environmental nonprofits. The *Reader's Digest* agrees, ranking the foundation among the twelve best charities in the United States. Jimmy Carter, nominally, chairs the foundation, but day-to-day business is run by someone in Juneau, the state's capital. The organization's board has included lawyers, authors, biologists, lodge owners, government employees, and accountants, some from Alaska, but some from Maine, Oregon, and California. What is the Alaska Conservation Foundation's business? They raise money, then give it away. They control almost seven million dollars, or about ten dollars for each person in Alaska. They give money to graduate students, they offer rapid response grants for emerging issues, they provide project support, and they offer operating budgets to help other nonprofits stay afloat. More than two hundred organizations have benefited from Alaska Conservation Foundation largesse.

The similarly named Alaska Conservation Alliance is a coalition of forty-five conservation groups and businesses representing something like forty thousand people. They build partnerships, they provide training, they offer technical support, and they lobby. Member organizations include Trustees for Alaska, Alaska Audubon, Friends of Potter Marsh, Earthjustice, and, surprisingly or not, Green Star.

There is, too, the Alaska Conservation Voters, the Alaska Forum for Environmental Responsibility, the Northern Alaska Environmental Center, the Southeast Alaska Conservation Council, the Alaska Marine Conservation Council, and the Alaska Wilderness League. This last one, despite its name, is located thousands of miles from Alaska, in the mountainless wilderness of Washington, D.C.

What sets Green Star apart from the rest of the state's nonprofits? They are more than a little unique in that they are not trying to protect the wilderness. They are not lobbying on behalf of wolves or bears or geese.

"We're a resource development state," Sean says, "and that's kind of shaped the people who live here. It's funny, living in such a pristine environment like Alaska. People come to Alaska thinking, 'Wow, everyone here must be fairly liberal, and everyone must put their recycling out.' But then you get here and you realize that you're in a resource development state."

Green Star started by recognizing the need to deal with trash. The average American throws away a bit over three pounds of trash a day, but the average Alaskan throws away about six pounds of trash a day. Sean saw the value of recycling. He knew, for example, that Americans throw away enough aluminum every year to rebuild the commercial air fleet twelve times. He knew, too, that the energy saved by recycling a single aluminum can could run a television for more than an hour. Recycling a glass bottle saves enough energy to run a light bulb for four hours. One thousand plastic milk jugs could retire to become a plastic park bench. By recycling paper, each person could save four trees each year—for all Alaskans, that would be more than two million trees, a small forest.

Sean liked recycling, but he took it one step further, seeing that waste included not just trash but air emissions and inefficient use of energy. And then he went one step further still, recognizing that eliminating waste could add to the bottom line. Reducing, reusing, and recycling, to Green Star, are tantamount to increasing profits.

I read him a few quotes from *Natural Capitalism*, a well-known book advocating environmental protection. The book describes garbage as "money being thrown away." Waste is "any human activity which absorbs resources but creates no value." It "represents money spent where the buyer gets no value." And the coup de grace: "The massive inefficiencies that are causing environmental degradation almost always cost more than the measures that would reverse them."

The quotes, Sean says, ring true.

"People I talk to," Sean tells me, "come from a variety of backgrounds. Certainly, though, a lot of people can see that waste is an unnecessary cost. And when it comes down to raising the topic with the big bosses, it's put more in terms of how much this will cost and what the savings will be. The easy sell is that it makes financial sense, especially within a five-year life span or a product life span, really a relatively short time."

I ask why a company would agree to join Green Star. "Three things pop into my head," Sean says. "One is to promote their companies as a caring member of society and part of the local community. A second one is to prevent pitfalls that can come up later—the legal counsel's perspective, keeping out of trouble. And the third one has to do with people in the organization. An organization is a group of people working together to fulfill whatever it is they do, and, when it comes right down to it, the people care. They care about the quality of the environment they live in."

Occasionally, Green Star membership helps with marketing. "Some organizations have given improved points or a stronger look at competing contractors who are Green Star members. I've heard that Alaska Railroad and the Anchorage School District ask if their contractors are Green Star members. We get calls from people bidding on contracts and looking for Green Star certification."

He pauses. "I think companies are looking at shorter and midterm payoffs," Sean tells me. "But a lot of employees are thinking about their grandchildren and future generations. The comments I get from some of the people I work with is that they will reduce waste for whatever company they work for. It is just their personal philosophy. They see long term, and they recognize that the way we are living on the planet right now is not necessarily sustainable."

A certain hotel in Anchorage, owned by one of Alaska's native corporations, applied for Green Star certification. The hotel, with thirty staff and 109 rooms, aspired to become known for ecotourism. They recycled office paper. They donated what customers left of the little plastic bottles of shampoo and conditioner and moisturizer to a local shelter. Unclaimed lost-and-found items went to a charity. Worn towels became rags. Worn sheets went to another charity. They used electronic faxes. When they printed, it was on double-sided pages. Around the hotel's back patio, native plants grew—paper birch and black spruce and poplar, which could, when attended to, look slightly better than scraggly. Newspapers were loaned out rather than distributed to every room, and at the end of the day they were recycled. The hotel sent outmoded computers and copiers to the Green Star electronics recycling event. Worn furniture went to the Anchorage Materials Exchange. Light came from ultra-efficient T-5 fluorescents. Guests had to call the front desk if they wanted to adjust the thermostat below sixty-eight or above seventy-two. The hotel might eventually put recycling bins in the rooms.

Among other things, Green Star advised them about a key-tag system. With key tags, a room key has to be inserted in a slot before room lights will come on. When the guest leaves for the day, the key goes with the guest, and the lights go out. According to Green Star, basing its numbers on data from another hotel, installing a key-tag system in just under three hundred rooms would cost about sixteen thousand dollars, and it would reduce energy costs by fourteen thousand dollars per year. Light bulbs would last longer, too, saving another two thousand dollars a year.

Welcome to Green Star.

Apolitical leanings aside, nonprofits cannot seem to stay out of trouble. The mantra might be "No good deed goes unpunished." Take the Nature Conservancy: big, rich, apolitical, loved by liberals and conservatives alike for its strategy of buying land for conservation. But organizations as big as the Nature Conservancy do not always act consistently. In May 2003, the *Washington Post* undertook what might be seen as something between reporting a feature story and entering a smear campaign. A front-page article started by acknowledging the conservancy's size, its three billion dollars in assets, and its practice of saving "precious places." But then came paragraph two: "Yet the Conservancy has logged forests, engineered a $64 million deal paving the way for opulent houses on fragile grasslands and drilled for natural gas under the last breeding ground of an endangered bird species." The authors criticized the Nature Conservancy for failed attempts to run a bed-and-breakfast, an oyster farm, and other businesses that led to twenty-four million dollars in debt. They did not like the fact that the conservancy occasionally made cash loans to its executives, failing to point out that this is a reasonably common practice in the world of business. The *Post* criticized the conservancy for not taking a stand on climate change and the Arctic National Wildlife Refuge, implying that no conservation group worth its name could or should be apolitical.

So what if the Nature Conservancy logged forests? Logging is a standard part of forest management; often, it makes sense to selectively log forests in ways that improve habitat quality for some species. Certain birds like openings in the canopy. Deer like plants that do not grow under the thick shadows of densely packed trees. Logging also yields profits that can be used to buy more land. Allegations of paving the way for opulent houses on fragile grasslands come from the Nature Conservancy's "conservation buyers" program, which allows steep discounts in exchange for conservation easements on land deeds. The deeds prevent construction of housing developments and, say, condominiums, but they may allow construction of a single home, which might just happen to be a mansion. Or, in the case of the sixty-four-million-dollar deal highlighted by the *Post*, a conservation buyers deal may allow a few houses to be built on land that otherwise might have been entirely built out and destroyed as viable habitat.

"There are trade-offs in conservation," Nature Conservancy president Steven McCormick was quoted as saying. "We make a judgment that less than 100 percent is acceptable."

The Nature Conservancy's decision to drill for gas under the habitat of the Attwater's prairie chicken may be less palatable than the conservation buyers approach to land management, but it can still be seen—at least at its conception, prior to spinning out of control—as a conservation compromise.

Attwater's prairie chicken, before its coastal grass habitat was whittled away to almost nothing, was once hunted as a game bird in Louisiana and Texas. By 1919, it was gone from Louisiana. Before World War II, fewer than nine thousand birds were left in Texas. The Endangered Species Act kicked in. In 1972, the Attwater's Prairie Chicken National Wildlife Refuge was established on several square miles of land originally purchased by another nonprofit, the World Wildlife Fund. There, the male chickens strut about, point their tails toward the sky, puff out orange air sacs in their necks, grunt, and stomp their feet, all in an effort to attract females. The females cannot resist such charms. Later, they lay up to a dozen eggs in nests scratched into the ground.

The Nature Conservancy saw the obvious: The Attwater's prairie chickens needed more land. In 1995, Mobil donated a piece of coastal prairie to the Nature Conservancy, calling the donation "the last, best hope of saving one of the world's most endangered species." The idea was to release captive-bred chickens on the land, establishing a viable population. Then things became increasingly peculiar. In 1999, the Nature Conservancy announced that it intended to drill. The *Washington Post* claims that things grew even worse when longtime Houston businessman Jack Schneider, acting on behalf of the Nature Conservancy, offered to buy oil and gas rights under nearby land owned by the Russell Sage Foundation of New York. The foundation discovered that Schneider was working for the conservancy and that he had reason to believe the foundation's mineral rights were worth far more than the twenty-six thousand dollars he had offered. The foundation in the end accused the conservancy of conspiracy. A Russell Sage Foundation lawyer had this to say: "What they did was basically try to steal our interest. . . . They lied. They lied. And they lied some more."

In the meantime, they also drilled. To be fair, they used directional drilling methods, coming under the chicken's habitat from the side. On the downside, though, construction of a pipeline prevented the timely release of the captive-bred birds. Late release meant the birds were more subject to predation, and five birds drowned in a pen during heavy rains.

What were they thinking at the Nature Conservancy? They were thinking that they would use natural gas profits to buy more land, ultimately protecting more chickens, and maybe a few other birds. After all, birds need money, too. But later, an internal Nature Conservancy communication stated that the management of this situation was "not consistent with our values." When asked why gag orders were part of a proposed lawsuit settlement, the Nature Conservancy's president explained, "We just didn't want people to talk about how . . . stupid we were."

It is fun to imagine Sean undertaking a Green Star assessment of the Nature Conservancy's drilling operations in Texas—not realistic, because Green Star does not operate in Texas, but nevertheless fun. Sean, with a clipboard, would wander around the site, sweating in the Texas sun, checking their light bulbs, poking around in their dumpsters, urging them to encourage employees to carpool or cycle.

Deborah Dasch, writing about another environmental nonprofit and an Alaskan mining operation, wrote: "At a major mine this year an attorney from the National Wildlife Fund met with the mine's environmental manager to see for himself the mine's environmental safeguards and reclamation plans. The lawyer said they had done a great job, but when the manager said, 'So we're okay then,' he replied, 'Oh no, we're still going to sue you. We don't want any mining in Alaska.'"

I once asked Sean how he felt about the more adversarial nonprofits. He said that he sometimes thought it might be fun to work for the more adversarial groups, and I immediately understood what he meant. Who would not want to approach the world as though only one issue matters? Who would not want to ditch all those social courtesies and the nuances of polite society and even on occasion the inconveniences of minor points of property law in a no-holds-barred effort to bring down corporate giants?

"I would like to see things done as green as possible," Sean says. "I would like to be able to ride my bike or ski to work everyday. Everybody has to decide on what level they are comfortable. Everybody has different motivations for different things. I used to ride my bike to work, but now I have a kid, and it's just not very convenient to put him in a bike trailer at zero degrees. So I sit in traffic. And then I see a road extension going in, and I'm torn. More roads make it easier to drive, so more people will drive instead of biking or taking the bus. But I'm part of the problem, too. I'm out there driving. So from a personal perspective, from my own values, maybe I'd like to see planning that encourages people to leave their cars at home, to take the bus, to bike. But from a Green Star perspective, we want the best efficiency and environmental management however things are going on. We catch more flies with sugar than vinegar. If we remain neutral, we can help big organizations make significant changes. Where, if we were radical, we would have diminishing returns. There is a place for that. There is a place for the radical groups, but I'm just not comfortable there. For me, from sort of a long-term view of how it happens, you've got to work within the confines of the real world. When I was younger I didn't see it that way."

He gives me a report on the building we had inspected. The report, for the most part, is a checklist, delivered electronically. After the checklist, it offers advice. With regard to light bulbs, "consider fluorescent lamps recycling. Lamps contain mercury, which is released to the atmosphere when lamps are broken." With regard to obsolete computers and worn furniture, "consider posting any usable but unwanted electronics, furnishings, or other items on the Alaska Materials Exchange." With regard to commuting, "investigate the possibility of offering staff People Mover bus passes or incentives to ride the bus." There are, in all, twenty-eight separate bits of advice. Alone, none of them are profound. In sum, though, there is a message here. It is the same message delivered by the National Environmental Policy Act: stop and think. It is a message that says, "Hey, before you throw that away, before you jump in your car to run down to the corner store, before you print that document, before you buy that product, stop and think about the waste."

In the Green Star office, I ask about their day-to-day habits. "We don't monitor our own trash," Sean says, "though we really should, just because we ask our members to do it. But I can tell you when we leave here at the end of the day there are maybe a couple of tissues and maybe a food wrapper, and that's about it."

CHAPTER SEVEN

BREATHING CLEAN AIR

Joe Kubsh's journey into air was direct: He has a PhD from work on smokestack technologies. Dave Foerter's journey was less direct: He eased into air from biology and toxicology, working at one point on the Chesapeake Bay Program, then moving into a transportation program that carried him naturally into smog. The two men work for sister trade associations, Joe for the Manufacturers of Emissions Controls Association and Dave for the Institute for Clean Air Companies. Both refer to their organizations by acronyms: MECA and ICAC. When they are not traveling—less than most of the time—they share the same no-frills suite of offices on the 1600 block of L Street in Washington, D.C., a few miles from the Environmental Protection Agency's headquarters and Capitol Hill. They also share a frustration: Both men feel they are seen by some as playing this system, as rigging the game so that big business will pay through the nose for air-pollution-control technologies supplied by the companies that support MECA and ICAC.

MECA member companies include the likes of Alstom Power, Forney Corporation, and Horiba Instruments. ICAC member companies include Applied Utility Systems, Babcock and Wilcox, and Mobotec USA. These are not household brands, but rather companies that invent, develop, manufacture, install, and service emission-control technologies, things like catalytic converters, particulate filters, bag houses, crankcase emission-control technologies, and scrubbers. They are modern day chimney sweeps, selling their products and services to sometimes reluctant automobile manufacturers and power plants and smelters.

"The power folks, the power and coal industry, very often they have a misconception that we're the bad guys," Dave says. "But we make coal clean. It's our member companies that can make coal a clean resource. We can make coal very clean."

In a separate conversation, Dave tells me that without the services of his member companies big industry would be stuck with one solution, or maybe two, rather than the plethora of solutions now on the market. "Our companies," he says, "give industry the flexibility of multiple technologies. What we do lets industry make choices about the most cost-effective approach to controlling emissions." In effect, they keep coal in business.

Joe elaborates on the same theme. "This market we work in," he says, "is entirely created by regulatory forces. Our members tailor pollution-control technology to meet the requirements of regulations."

When asked if MECA's member companies expect him to lobby for specific legislation, Joe is emphatic that they do not. "Our charter is to serve as a clearinghouse for technical information," he says. "We describe the capabilities that have been developed and are out there. We were formed as an interface between the companies and the regulatory community. We're not a lobbying organization. We're a technical information source for the industry."

"Our membership," he tells me, "is not open to the original equipment manufacturers. We're only open to the class of developers that sell into that marketplace." Companies that make cars, trucks, trains, and off-road vehicles cannot join MECA.

In a separate conversation, Dave says the same thing about ICAC. "ICAC," he tells me, "is the only purely air-pollution-control technology association for stationary sources. There are others that you can find on line for advertising and this sort of thing, but they don't help these markets move forward. Once these markets are there, they try to facilitate interactions between customers and vendors, but they don't actually try to help make sure we have the sound public policy that's needed to create the markets. We're purists in that regard. We don't have power companies in our midst. We don't have customers in our midst. We put out information that will help people understand how to do things. We put out technical guidance. We say, 'Here's what the technology can do. Here are different products that are completed.'"

Referring to the companies that buy clean air technologies, Joe says, "We're often on different sides of the issues. Some of the auto companies until recently generally fought tougher regulatory programs in this country and around the world. They saw it as a cost. When it comes to regulatory issues, more often than not we're on different sides."

Grant Ferrier might say that these companies did not want to internalize the cost of emissions. They preferred, instead, to socialize costs and privatize profits.

Here is a question one might ask: In this overpopulated world, in this era of environmental expectations, would companies that emit air pollutants even be in business without the devices sold by MECA and ICAC member companies? Would they be able to operate without air pollution regulations? The answer: If they were in business, they would be spending tremendous sums of money scrubbing the soot from their factory windows, finding substitutes for workers laid up with bad coughs and asthma, and training new workers to replace those who dropped dead before retirement. And, too, their customers would have dirty windows and chronic coughs. By one reckoning, the estimated value of air pollution control in the United States is just over a trillion dollars a year, driven largely by improved health and increased life expectancy. In contrast, the estimated annual cost of air pollution control is a paltry twenty-seven billion dollars. If the game is rigged so that emitters pay through the nose to the companies that make up MECA and ICAC, so be it.

Average adult lung capacity is about five quarts. A full breath holds about one quart of oxygen, four quarts of nitrogen, and a few tablespoons of carbon dioxide. The oxygen keeps us alive. The carbon dioxide comes from our own exhalation and from burning wood or coal or oil or natural gas. The nitrogen comes along for the ride, filling space but doing neither harm nor good. Also along for the ride, but considerably less harmless, is a pinch or less of benzene, beryllium, mercury, vinyl chloride, dioxin, lead, mercury, and arsenic—all members of the nearly two hundred toxins believed by the Environmental Protection Agency to cause cancer, birth defects, and decreased intelligence. Mixed with these are chlorofluorocarbons, halon, and methyl chloroform, chemicals that, higher in the atmosphere, cause holes to form in the stratospheric ozone layer, letting in excessive ultraviolet light that can, among other things, cause skin cancer. There is also sulfur dioxide, which, when it contacts fluid in the lungs, becomes sulfuric acid. There is carbon monoxide. And there is particulate matter—dust from roads and power plants and car exhaust—known in the business as PM10s and PM2.5s, signifying the 10- and 2.5-micron diameter of particles. To call these particles tiny would be to exaggerate their bulk. For perspective, one inch is 25,400 microns long. Because they are small, they can be dangerous to human health. They find their way deep into lungs. New evidence suggests that these very small particulates are even more dangerous than was once believed.

But there is nothing new about air pollution itself. Seneca complained about it in ancient Rome. In 1257, the Queen of England left Nottingham because of the smoke. In 1285, King Edward I, fed up with London's air, established what might have been the world's first air pollution commission. Twelve years later burning of coal was outlawed in London, albeit unsuccessfully; people preferred breaking the law to shivering. From the *London Times* in 1812: "For the greater part of the day it was impossible to read or write at a window without artificial light."

The United States suffered, too. A visitor to Pittsburgh around 1820 described "a cloud which almost amounts to night . . . with the appearance of gloom and melancholy." Reportedly, fashion-conscious residents wore black because black clothes do not show soot.

Laws evolved. Chicago and Cincinnati enacted clean air laws in 1881. At the turn of the century, the Bureau of Mines created an Office of Air Pollution. In 1955, Congress passed the Air Pollution Control Act, offering five million dollars a year for research into the extent of the air pollution problem. In 1960, the act was amended to allow continued funding. In 1962, it was amended again, this time to bring the Surgeon General's office on board, further recognizing that bad air posed a health risk. Eight years later, the Clean Air Act of 1963—the first piece of federal legislation to use the phrase "Clean Air"—gave ninety-five million dollars over three years to state and local governments for research and establishment of air-pollution-control agencies. The Clean Air Act was amended in 1965, 1966, 1967, and 1969, setting standards for automobile emissions, establishing air quality regions, and authorizing expenditures for research on low-emission fuels and vehicles. The 1967 amendments set a schedule for State Implementation Plans, or SIPs, which would, over time, become a key part of clean air legislation.

In 1970, the first Earth Day coincided with discussions of further amendments to the Clean Air Act. In Washington that year, it was hotter than normal, making the air seem heavy. The amendments, under the leadership of Senators Muskie, Eagleton, and Baker, focused on public health, with the natural environment taking a back seat. Importantly, and despite a coalition of labor and environmental groups calling for the banning of the internal combustion engine, the amendments recognized that technology could fix this problem. The amendments were not about abandoning technological progress, but about taking that progress one step further.

Senator Muskie has since called the amendments, commonly referred to as the Clean Air Act of 1970, "the most far-reaching piece of social legislation in American history." The amendments set National Ambient Air Quality Standards, or NAAQS, to protect public health and well-being. They established New Source Performance Standards to regulate new power plants and smelters and refineries. They set standards for cars. They used deadlines to drive performance and to force accountability. They required companies to disclose the contents of their emissions. They coughed up more money for research, including thirty million dollars for research into noise pollution alone. And here was something interesting: The amendments gave citizens the right to take legal action against the government or any other organization that violated the emissions standards, essentially deputizing the entire nation, saying, "It's your air, so keep an eye on it."

On top of all this, the Clean Air Act of 1970 identified what it called "criteria pollutants." The criteria pollutants were the common pollutants: ozone, volatile organic compounds, nitrogen dioxide, carbon monoxide, particulate matter, sulfur dioxide, and lead. Ozone, in this case, is ground-level ozone, which is harmful to humans and the environment, as opposed to stratospheric ozone, high in the atmosphere, which blocks unwanted ultraviolet light. The ground-level ozone forms from the chemical reaction of other priority pollutants and can form well downwind from the source of these pollutants. It is the main ingredient of smog; it causes asthma, stuffy noses, and reduced resistance to colds and other infections. In short, it screws up the body's breathing machinery. Volatile organic compounds—things like benzene, toluene, methylene chloride, and methyl chloroform—are among the pollutants that react to become smog. They come from paints, solvents, and glues, or from burning of natural gas, gasoline, and diesel fuel. In addition to forming smog, they can cause cancer. Nitrogen dioxide, another of the smog precursors, comes from burning gasoline and other fossil fuels. It damages the lungs and breathing passages. Carbon monoxide binds to hemoglobin in the blood, reducing the blood's ability to carry oxygen. Like most of the other criteria pollutants, carbon monoxide comes from burning gasoline and other fossil fuels, especially when those fuels are burned inefficiently—when, for example, someone in Anchorage cold-starts their car at ten below. Particulate matter comes from burning wood, diesel, and other fuels, but it also comes from the plowing of fields and other activities that raise dust into the air. It reduces visibility and discolors clothes and other property. More importantly, it causes nose and throat irritation, lung damage, and bronchitis. Sulfur dioxide comes from burning high-sulfur coal and diesel fuel. It causes breathing problems. In contact with water, it becomes sulfuric acid, the famous acid rain that kills lakes and forests and eats away at statues and melts the faces of gargoyles on Gothic cathedrals. And there is lead, which, when the Clean Air Act of 1970 was passed, was still intentionally added to gasoline to reduce engine knocking. It went from fuel tank to engine to tailpipe to the air to the lungs. It comes, too, from some paints and smelters. Breathe enough of it, and it damages the brain.

In addition to the criteria pollutants, the act listed 189 substances as "hazardous air pollutants." They are less ubiquitous than the criteria pollutants, but locally, where they are released, they can be unhealthy. In concentrated forms or over long periods, they can be deadly. For the most part, they are chemicals of which most people have not heard, things like biphenyl, ethyl acrylate, hydrazine, and quinoline. But there are also more widely recognized chemicals, like chlorine and hydrochloric acid. These are substances known to cause cancer, birth defects, and environmental effects. They come from burning fossil fuels, from paint, from household products, from dry cleaners.

The Clean Air Act went through another major amendment twenty years later, in 1990. In nine sections covered in an eight-hundred-page document, the 1990 Clean Air Act required the Environmental Protection Agency to issue 175 new regulations, prepare thirty guidance documents, and write twenty-two reports. It started fifty-three new research projects. It mandated gasoline reformulation in the nation's dirtiest cities. It allowed for the withholding of highway grants to states that did not cooperate. It mandated the use of "Maximum Achievable Control Technologies for the Hazardous Air Pollutants." It mandated reduction of sulfur oxide emissions by almost nine million tons per year. It banned use of chemicals like halon that would erode the stratospheric ozone layer. It authorized a fee system for emissions that would lead to acid rain.

Here is another way to summarize the history of the Clean Air Act: A new regulation is proposed, industry objects because implementation will be too expensive, a compromise is reached, costs of implementing regulations are lower than anyone expected, and no one would dream of going back to preregulation conditions.

An ICAC press release puts it another way: "It is well to remember the preeminent lesson of our Nation's 27-year history under the Clean Air Act: Actual compliance costs turn out to be much lower than the costs predicted at the outset of a regulatory action because the regulated community, markets, and control technology suppliers are smarter and more efficient at reducing costs than forecasters predict." An example: In the late 1980s, industry believed that proposed reductions in sulfur dioxide emissions would cost fifteen hundred dollars per ton, but after implementation they discovered that the actual cost was less than a hundred dollars per ton. Another example: Industry feared that lead requirements in the 1990 Clean Air Act would cost between about fifty and a hundred billion dollars a year, but true costs turned out to be closer to twenty-five billion dollars a year.

Yet another way to summarize the history of the Clean Air Act: By 2000, after three decades of serious clean air legislation, carbon monoxide levels had decreased by 31 percent, sulfur dioxide levels had decreased by 27 percent, the smallest particulates had decreased by 71 percent, volatile organic carbons had decreased by 42 percent, and lead had decreased by 98 percent. All of these decreases occurred during a period of population and economic growth. According to a government study, for every dollar spent on compliance with the Clean Air Act between 1970 and 1990, twenty dollars were saved through improved health and environmental conditions. In 1990 alone, the Clean Air Act saved seventy thousand lives and prevented fifteen million respiratory disorders, including 13 million cases of high blood pressure, eighteen thousand heart attacks, ten thousand strokes, and nearly a 150,000 cases of respiratory complaints.

And one more way to summarize the Clean Air Act: Today, despite the Clean Air Act and its various amendments, about ninety million Americans, roughly one in three, live in what the Environmental Protection Agency calls "nonattainment areas." Nonattainment areas fail to meet the government's health standards for air quality.

Senator Muskie, in a written statement about the evolution of the Clean Air Act, had this to say: "One shudders to contemplate the magnitude of the problem with which we would be confronted had we neglected to address these conventional pollutants 20 years ago."

In his office at ICAC, Dave Foerter talks to me about his organization's interest in what he refers to as "stationary sources"—power plants and smelters and refineries and various industries as opposed to buses and trains and trucks.

"The power companies want certainty," he tells me. "They want to know what they will have to do. The only way they will get that is to go back to the public health record. Eventually, they'll have to do something to satisfy health concerns, and that is where the certainty is. Every time a new initiative comes forward, they fight it, and that increases uncertainty. They want certainty, but they don't like the reality of certainty. If companies would do the right thing, no one would ever ask for regulation. It just isn't happening. Sometimes they'll go beyond what they need to do, but generally they don't. These are corporate decisions. This is money. That's what drives these guys."

ICAC, with a three-person staff, runs as lean as Green Star, yet they keep their industry informed about new regulations and they keep regulators and policy makers tuned in to new technologies.

"ICAC member business is worth something like ten billion dollars a year," he says. "It's a big industry. Just the multipollutant-control programs created something like three hundred thousand new jobs. This was in 2002. These are good jobs, high technology jobs, not flipping burgers."

His budget, for all of ICAC, is half a million dollars a year. "We have nineteen pages of technical comments on mercury that are due tomorrow. They're not from me; they're from experts in the member companies. I am able to harvest the members' expertise. If I knew all this stuff myself, I'd have a head so big—" He finishes the sentence by holding his hands up, as if supporting a beach ball on top of his shoulders. The half-million-dollar budget at ICAC mobilizes millions of dollars in effort from the member companies.

I ask about what is going on now, about where his member companies see the greatest opportunities.

"The biggest issue for the power industry is PM2.5s." Dave is referring to the smaller-than-tiny particles, the particulate matter way too small to notice, but nevertheless deadly. By some estimates, PM2.5s kill 135,000 people a year.

"There's a lot of stuff in there," he says. "If you control PM2.5s, you control mercury, nickel, other stuff. By controlling PM2.5s you get a free ride on other pollutants." You get for air what Gordon Austin, for water, would call incidental treatment.

Baghouses are one way of controlling particulate matter. The baghouse is old technology. It is what the name suggests: a house full of cloth bags. Gas from smokestacks passes through the bags, and dust from the gas collects in the bags. In what in itself is another form of incidental treatment, the dust then captures the smaller particles, the PM2.5s. But when thinking of baghouses, think big. A baghouse can be almost as long as a football field, with more than ten thousand bags, each bag in itself big enough to hold a school bus. The size and expense of baghouses have been deterrents, but clean air legislation encourages their use. In 1975 baghouses were rare in American power plants, but by 1989 fifty-seven baghouses were in use. Today, more than half of new power plants use baghouses.

Our conversation wanders to what is often called "technology forcing," in which regulations are passed that cannot be met with currently available technology. Companies have no choice but to pursue new technologies.

"The phrase 'technology forcing' really came from California's low-emission vehicle program," Dave says. "It tends to have a negative connotation with it. 'Forcing' sounds bad. But it's really a very positive thing. If you want to continue to make progress, something has to force that progress to occur. 'Technology promoting' would be better wording."

Sulfur dioxide from stationary sources—the principle cause of acid rain—could be the poster child for technology forcing. The forcing mechanism has been the cap, or the total emissions that can be released in a region. "As the cap becomes smaller," Dave says, "or as more companies need to emit pollutants, it forces new technologies, cleaner technologies."

The 1990 Clean Air Act put a cap in place for power plants, reducing their total sulfur dioxide emissions to half of what they had released ten years earlier—half, that is, of the almost eighteen million tons that had been released in 1980. In the first phase of the new requirements, 110 plants had to reduce their emissions, but reducing the amount of power they produced was not a realistic option. The power industry was forced to figure something out. As a tool, they were allowed to trade. People talk of "cap and trade" policies. If one plant reduced emissions by more than half, it could sell its surplus allowance to another plant to meet the cap. And they could be creative. Many plants resorted to what is sometimes called fuel switching, changing from high-sulfur coal to natural gas or oil. Other plants switched coal types, abandoning the higher-sulfur coal from mines in the eastern states for lower-sulfur coal from mines in the West. Still others used improved technologies, the kinds of things developed and sold by ICAC's members. Example: They inject a slurry of limestone into the flue gas, where it reacts to become carbon dioxide and calcium sulfate; then they release the carbon dioxide, an unregulated emission, into the atmosphere and collect the calcium sulfate. Another name for calcium sulfate is gypsum, the stuff inside the sheetrock walls of most American homes. The process itself is not new, but its improved efficiency is. With new technologies, 90 percent of the sulfur dioxide can be removed and converted to gypsum at an estimated cost of about three hundred dollars per ton.

"Recently," Dave says, "the emphasis has shifted from cap and trade to just plain trading. But you need to keep your eye on the cap. That's the goal—keeping the cap down. The idea of trading is to give industry flexibility. Okay, we have the cap, but we have to give industry something, so we give them trading. But the cap is the budget. If someone asks me, 'Are you an advocate of trading,' I would say, 'I'm an advocate of good trading.'"

Others feel more strongly. A nonprofit organization called the Clean Air Trust calls trading programs "a dismal failure." Others see ethical problems with the very concept of trading, which, from one perspective, allows companies to buy the right to discharge poisonous gas into the atmosphere. On the other side of the fence, the Department of Energy has said that trading "capitalizes on the power of the marketplace to reduce sulfur dioxide emissions in the most cost effective manner."

"Trading," Dave adds, "creates a competitive atmosphere. Our members are very competitive. Members may work together on some issues, but they are very competitive."

And here is another result: Forcing lower sulfur dioxide emissions in itself reduced the acid rain problem, but this was not expected to have a real effect on human health. Unexpectedly, though, reducing sulfur dioxide led automatically to reduced particulate releases and a subsequent positive effect on health—again, incidental treatment.

Cost to the consumer of removing sulfur dioxide is about twelve dollars per person.

ICAC maintains a collection of success stories. In Jersey City, a member company provided what is described as "a three-pass electrostatic precipitator with microprocessor-controlled intelligent power supplies" to a cafeteria, reducing particulate emissions by 95 percent. In northern California, a member company provided catalytic converters that removed nitrogen oxides and carbon monoxide from a power plant's exhaust stream. Another member company helped a foil manufacturer in Michigan with a system that took volatile organic carbons from the exhaust stream and burned them to run equipment, dropping energy costs 75 percent and reducing emissions of organic compounds. Another worked with an incinerator in Belgium to install a filter laminated to a catalyst, capturing dust while simultaneously breaking down dioxin. Another removed sulfur dioxide and sulfuric acid mist from a mine's emissions in Virginia. Still another controlled particulates from a plant that burned waste wood in Long Beach, California.

"If the emission can be controlled for a reasonable cost," Dave says, "just do it. Just do it. Now what is the reasonable cost? For reducing costs of oxides of nitrogen, a reasonable cost might be a couple of thousand dollars a ton. But reducing hazardous air pollutants shouldn't be to the same standard. It should probably be tens of thousands of dollars per pound because of the impact on health."

We return to the topic of success stories. "There was a facility with hydrogen chloride," Dave says. "Hydrogen chloride was their biggest emission. The regulatory process required them to control the hydrogen chloride. What they wound up doing was selling the stuff. They got a net profit. Instead of having a problem with spewing this stuff into the environment, now they have a profit."

Succinctly, he describes the industry he represents: "Give us a problem, and we'll find a solution."

Joe Kubsh and his four colleagues at MECA work down the hall from Dave. Their activities and philosophies mirror those of ICAC, but while ICAC works on stationary sources, MECA works on mobile sources: cars, buses, trucks, trains, and, recently, lawn mowers. The division—stationary versus mobile—may not at first make sense since the pollutants produced substantially overlap. But more closely examined, the division makes sense. Clean air laws have always distinguished between the family car and the big smokestack industries. And, too, the technologies used to control pollution from mobile sources are different than those used for stationary sources. MECA's forty-two member companies are not selling the same products that ICAC's members sell.

MECA started in 1976, about the time that catalytic converters were first put on cars in the United States. Catalysts and associated technologies are the bread and butter of MECA member companies. Catalytic converters sit in the exhaust system, just upstream of the tailpipe, promoting chemical reactions that otherwise would not occur. Catalytic converters are found on all but the oldest cars on America's roads. They are part of the full suite of emissions controls on modern cars. Exhaust passes over a mix of platinum, rhodium, and palladium, and nitrogen oxides and carbon monoxide otherwise bound for the atmosphere become harmless gases like nitrogen and oxygen.

"Most drivers are familiar with emissions control because of required car inspections," Joe tells me. He estimates that the consumer pays between five hundred and a thousand dollars for emissions control technology on a typical new car, including the catalytic converter, fuel injection, computer chips that manage engine temperatures, and the various other innovations that contribute to automobile emissions controls. In exchange, the consumer gets cleaner air. Relative to a car of 1970, a new car emits 96 percent less carbon monoxide, 95 percent less nitrogen oxides, and 98 percent less hydrocarbons. Because lead destroys catalysts, requirements for catalysts contributed to the phase out of lead. Altogether, something like one and a half tons of emissions per car each year no longer reach the atmosphere.

Just how big a deal is all this? The Society of Automotive Engineers selected the catalytic converter, fuel injection, and electronic engine controls—all developed to reduce emissions—as among the industry's ten greatest achievements in its first hundred years.

"The leadership position," Joe says, "is California. Because of their unique air quality problems, they've traditionally been at the forefront of regulatory programs. They put the best available technology on any source they can think of, including automobiles. They have the toughest emission-control regulations for light-duty vehicles in the world. They have almost zero emissions from some passenger vehicles. What's coming out of the tailpipe is cleaner than the combustion air that is going into the engine. In some ways you can conceive of driving the air clean. If the whole fleet was zero emission, it could clean up the ground-level ozone problem in the Los Angeles basin."

Here, when Joe talks about emissions and clean air, he is talking about particulates, carbon monoxide, nitrogen oxides, sulfur dioxide, and volatile organic carbons, but he is not talking about carbon dioxide. Carbon dioxide in the United States is not regulated. Although it is the number one cause of global warming, and although the nation produces a quarter of the global total, it is not, for now, considered a pollutant in the United States. If the United States agrees to the Kyoto Protocol or to some similar arrangement designed to stabilize or decrease the amount of carbon dioxide in the atmosphere, this could change.

Joe's member companies depend on regulations. "If the regulations aren't tight enough," he says, "people won't use the product." For now, regulations do not regulate carbon dioxide, and emissions controls do not control it.

MECA has grown in the last ten years. "Most of that growth," Joe says, "can be attributed to interest in technologies to clean up diesel engines. Maybe 70 percent of our activities right now are focused on clean diesels—mostly trucks but also moving into agricultural and locomotive applications. The latest suite of regulations calls for 90 percent reduction in particulates and nitrogen oxide."

Recently, the city of Washington bought two hundred clean-diesel buses. The low-sulfur fuel that they burn costs a few cents more per gallon than other diesel. Aside from inconspicuous labels, riders have no way of knowing they are on clean-diesel buses, unless they notice that the older buses continue to belch out plumes of soot.

Because of all the diesels already on the road, and because diesels tend to run for many years without replacement, Joe sees a need to retrofit existing vehicles. Particulate filters, he feels, are especially important. "There are limitations to what you can do with the black smoke coming from diesels that you see driving down the road," he says. "The filters require specific temperatures to burn off the soot." A retrofit costs around ten thousand dollars, and the filters need to be serviced, increasing maintenance costs.

Despite his belief that regulations drive the market, he acknowledges voluntary progress and market forces that move some manufacturers beyond current requirements. "The auto companies recently have a lot more green image about them than they've ever had. Some are very aggressive—companies like Toyota and Honda. Some of the American companies are narrowing the gap." He sees some of this as a response to the marketplace; in most areas, there are waiting lists for hybrid cars that run on gasoline and electricity, the kind that Mike Rolband owns. Joe is less optimistic about development of hydrogen-powered cars. "There's a lot of popular press around hydrogen vehicles," he says, "but my own personal opinion is that we are fifteen years away, maybe further, before we see large numbers of hydrogen-fuel-cell vehicles on the marketplace. Some people see 2050 as a more reasonable date."

But there are other green innovations. Example: Some car manufacturers have radiators coated with catalysts. "The catalysts destroy ground-level ozone as you drive," Joe says. "They clean the air as you drive. In California, there are credits that can be gained for implementing this technology. But some car manufacturers—Volvo is an example—see this as a marketing tool." An advertisement might show joggers behind a car, sucking in the clean air.

While marketing plays a role, the regulations give credit to car manufacturers that drive ahead of the curve. "Credits are usually based on early introduction," Joe says. "Regulations might start in 2007, but if you bring in vehicles ahead of that time, you generate some credit to maybe influence how fast you change over the rest of your fleet. Or you might be able to sell credits to your competitors who aren't going up the technology curve fast enough. It's a complicated credit system. I couldn't begin to explain all the nuances to you. But typically, it's within a sector. You're not allowed to trade heavy-duty vehicles for light-duty vehicles."

After a pause, he adds that the credit systems are not a huge driver of improvements. "They're just a way," he says, "for the

EPA to provide flexibility." They are just another manifestation of cap and trade, and while they may not drive improvements, the flexibility they allow may make new requirements more palatable to the very powerful automobile manufacturing sector.

I head to Annapolis to meet John Sherwell, who works for Maryland's Power Plant Research Program. He is highly regarded by both Dave and Joe. In traffic, I look for a retrofitted bus, but all of the buses I see spit black smoke when they accelerate. Somewhere, I take a wrong turn. This is not as troubling as one might suspect; I am accustomed to being lost in Washington. After twenty minutes of detours and one-way streets and roads dead-ending into stolid government buildings, I spot a hybrid car bearing the license plate "HOV." Incentives work. Like Mike Rolband, this person bought his hybrid not so much to save the planet as to save time by taking advantage of rules that allow hybrid vehicles in the faster-moving high-occupancy-vehicle lanes. For lack of a better option, I follow the hybrid. And, surprisingly, it turns east onto the highway to Annapolis, my own destination, where I hope to learn something about the pollutants that result from generating Washington's electricity. For forty minutes, I drive amidst diesel trucks that belch particulates, my rental car contributing its own carbon dioxide and nitrogen oxides and particulates, then park outside of a government building in Maryland. I find John Sherwell, and we spend the next two hours talking, sitting at a picnic table next to a pond just outside of his office. He tells me about starting life in South Africa, going to London for a PhD in the 1970s, and returning to his home.

"All of my research was focused toward air quality engineering," he says. "The clean air business back then was all about acid rain. In South Africa I worked for the government. They had huge coal deposits, and there were concerns about acid deposition. These were rather bleak times in South Africa. The apartheid government was particularly vicious at that time. People were disappearing. Friends were in jail or in exile. One of the things that precipitated me leaving was when I was asked to look at heavy gas dispersion. If they were going to use chlorine or something like that, as a weapon, they wanted to know how would it spread."

Rather than contributing to the government's knowledge about chemical warfare, John accepted a postdoctoral research appointment at Emory University in Atlanta, then moved to the University of Melbourne in Australia, then back to Emory, and finally to a consulting company in Texas. "The environmental industry was still new in the eighties," he says. "There was tremendous growth in the industry. Businesses needed support in understanding the regulations. Government needed support in writing the regulations. Consultants were right in the middle. Things peaked in the early nineties. The huge growth ended. Consulting companies started acquiring one another. It seemed like a good time to get out."

Grant Ferrier had painted a variation of the same picture. At one time, Grant had said that anyone with a business card could get work, but that this had changed. Competition became ferocious.

John landed at the Power Plant Research Program, by then already more than ten years old. "The program," he says, "came into being in the early seventies. Baltimore Gas and Electric was planning to build this nuclear power plant—the Calvert Cliffs Plant—as part of its suite of power capacity. It would be the rate payers who paid for the cost of the new facility, so the Public

Service Commission had to say, 'Yes, this is a wise use of the public's money.' Baltimore Gas and Electric showed up with its experts and the opponents showed up with their experts, and, of course, they were totally opposite, leaving the Public Service Commission stuck. So the legislature created the Power Plant Research Program. The idea was that we would be an impartial scientific body that would advise on issues related to electric power generation."

There are just over thirty power plants scattered across the state. More than half of the power comes from coal. About a quarter comes from the state's only nuclear plant at Calvert Cliffs. Most of the rest comes from oil and gas, but a small percentage comes from hydroelectric dams. A smidgen comes from landfill gas, piped up from underground garbage and burned in systems similar to that of the OII Superfund site in Los Angeles. At the Eastern Correctional Facility, wood provides electricity. A proposed plant might eventually burn chicken droppings.

From a Power Plant Research Program public relations release: "We need power plants and transmission lines to support the basic needs of modern society. At the same time, we must continually ask: What impacts do power plants have on the environment? Are the impacts significant? What are the costs to minimize the impacts?"

Plants that run on fossil fuels do not necessarily trigger the National Environmental Policy Act's requirement for environmental impact statements, but Maryland requires them, so the Power Plant Research Program writes what amount to environmental impact statements. The program also conducts research, trying to fill gaps in understanding about the kinds of environmental impacts that can come from power generation. They collect data. They write reports. They fund graduate students. They publish papers in technical journals with titles like *Atmospheric Environment* and the *Journal of the Air and Waste Management Association.*

"The kind of research we do," John says, "is not constrained by the regulations. It hasn't always made us friends." At first I think he is implying that his work frustrates industry, but he corrects me. "The regulators don't like people rocking the boat either," he says. "We tend to be well received by industry because we're not regulators. What we find is what we write. We just provide facts."

The research, the environmental impact statements, the rocking of the boat—all of this falls on the shoulders of five state employees, including John. These five do their own work and manage contractors. "We use contractor staff as though they are an extension of our own staff," John says. "It makes for quite efficient government." They spend ten million dollars a year. The money comes from monthly electric bills, working out to something like twenty-five cents per month for a typical bill.

Maryland, like other states, has deregulated the power industry. In the past, power generation was a controlled monopoly. Single companies would provide power for a region, and rates were set with government approval. The companies worked on what might be thought of as a cost-plus basis. Some felt that competition should exist, that the free market would lead to efficiencies and lower prices. So, slowly, the industry has been deregulated. Anyone can build a power plant and contribute to the grid. Overnight, incentives for what had been called "demand side management" ended. In demand side management, a power company actively encouraged consumers to use less electricity, especially during periods of peak demand. They did this through pricing practices that charged more for electricity during peak hours and through conservation programs that encouraged customers to install efficient lighting, heating, and air-conditioning systems. This would be the equivalent of an oil company discouraging people from driving or a liquor store supporting sobriety. But because of the absence of competition and a pricing structure set by the government, the regulated monopoly's profits could be higher if it avoided building a new power plant. A regulated monopoly might be more likely to propose voluntary emissions reductions, knowing that it would be compensated on a cost-plus basis.

Would consumers buy green power in a deregulated market? "Every now and then," John says, "you see green advertising. There are all these big hydrodams on the Susquehanna. After years and years of trying, we got them to put in fish ladders and fish lifts so that fish could spawn all the way up to New York. The Susquehanna dam on the river is a huge dam—a thousand megawatts. That was the end of the spawning run for shad. It took fifteen years to negotiate construction of these fish ladders. As soon as it was finished, there were all these ads about what a wonderful job they had done."

But, in general, power companies do not seem to see a competitive advantage in advertising environmental performance. "Among the major vendors," John says, "they haven't taken the environmental shtick. It may happen as we get further into deregulation, but it's not something we're seeing right now."

Concerns about the impact of deregulation on the environment led to new legislation. State law requires power plants to track and report changes in emissions resulting from deregulation. It requires the state to consider what is sometimes called a "renewable portfolio standard"—a system that forces the state's energy mix to include a certain percentage of renewable energy.

The state, which buys eighty million dollars worth of electricity each year, hopes to buy 6 percent from green producers. The law also requires power companies to explain the emissions resulting from generation. To this end, the consumer's monthly power bill contains the following wording: "Carbon dioxide is a 'greenhouse' gas, which may contribute to global climate change. Sulfur dioxide and nitrogen oxides released into the atmosphere react to form acid rain. Nitrogen oxides also react to form ground level ozone, an unhealthful component of 'smog.'" Above this warning, a table lists emissions released for each megawatt-hour of electricity generated. A household might use as much as one megawatt-hour per month, with an emissions cost of ten pounds of sulfur dioxide, three pounds of nitrogen oxides, and up to a ton of carbon dioxide.

From a state report on electrical power in Maryland: "No electric generation technology, however, is completely free of environmental impacts. For example, wind generation entails land use impacts, can result in the deaths of migrating birds and raptors, and can have visual impacts on adjacent communities."

To describe wind energy as scarce would be to overestimate its importance in Maryland. It would be more accurate to describe wind power as not worth mentioning. In a pie chart that breaks down Maryland power sources, wind is lost in the pie slice labeled "Other." Lost in the same pie slice are solar, tidal, and waste-burning generators.

The Power Plant Research Program's participation with alternative energy has been frustrating. "These wind power projects that we did were difficult," John says. The wind towers were 300 feet tall, with propellers 120 feet long. "The green community was quite opposed to them. There were some of the local people who just didn't want to see the landscape changed. There were some who were genuinely concerned about bird strikes. The wind machines get bigger and bigger. Size restriction is the size of what can be moved on land. Offshore, as the Europeans are doing, there's less size limitation."

In Denmark, the world's leader in wind power, 9 percent of the power comes from wind. These windmills kill thirty thousand Danish birds. In the United States, where less than a tenth of one percent of power comes from wind, seventy thousand birds die. But there is an issue of context here. Danish traffic kills a million birds a year. American traffic kills nearly sixty million birds. Cats eat fifty-five million birds in Britain alone. Plate-glass windows nail another ninety-seven million.

Costs of green power are more than just birds. Some forms of green power just cannot compete, especially when the cost of emissions is low or—in the case of carbon dioxide emissions in the United States—free. A kilowatt hour of electricity from a solar cell costs something like twenty-five cents, two to five times the cost from coal-fired plants and twice that of nuclear plants. And there is the issue of feasibility. Solar cells work poorly in northern latitudes. Wind power does not work when the air is still. Geothermal power, generated from naturally occurring steam trapped in the earth, is not an option on the East Coast. But there is also the issue of momentum. People accept what they know best. They do not want to invest in new technologies. They want to keep using what they have already built. Wind, although it produces electricity for about the same cost as coal, requires a substantial investment in new equipment. Two Stanford engineers, writing in the prestigious academic journal *Science*, suggested that construction of 225,000 wind turbines could displace two-thirds of the coal-fired plants in the United States. The cost is a startling $338 billion dollars to build and $4 billion a year in maintenance, not including the inevitable cost overruns. And this does not include the cost of shutting down the existing coal-burning plants that the wind generators would replace.

Over the past few years, Americans have grown excited about energy from hydrogen. But hydrogen's future, if it has one, is in storing energy. "I think the hydrogen economy is more of a political thing than a reality thing," John says. "You've got to get the hydrogen somewhere. You could use the energy from coal to make hydrogen from water, then inject the carbon dioxide emissions underground. You would essentially be using the hydrogen as a battery, but there are inefficiencies in the process."

On the other hand, he sees promise in nuclear: "If you want to get a zero-emission electron in your house," he tells me, "get all your power from nuclear. Twenty years from now, I could see nuclear coming back. There is a lot of rethinking in the nuclear business. Nuclear plants present a manageable risk no more dangerous than, say, a refinery."

For now, though, John sees coal as maintaining the lead. "Current federal policy is to support coal," he says. "There's going to be a push on clean coal."

Skepticism permeates John's comments about alternative energy technologies. But this skepticism is offset by his personal sense of responsibility. When he reads the label on his power bill, he understands the environmental costs of a megawatt-hour. Like Joe Kubsh and Dave Foerter, he knows that particulates kill more than a hundred thousand people a year, that sulfur dioxide turns rain into acid, and that five of the state's ten largest emitters of substances regulated as toxic chemicals are power plants. He knows that a single plant in Maryland can release eighty thousand tons of sulfur dioxide and twenty-four thousand tons of nitrous oxides into his air each year. He knows that on average every house in Maryland is, in effect, dumping more than a ton of carbon dioxide into a warming atmosphere every year.

But he can think in terms of zero-emission electrons flowing into his house, at least as a possibility. "I would be willing to buy green power," he says, "even on my state salary. It's something that needs to be supported."

CHAPTER EIGHT

GREEN CARPET

Ray Anderson may be the chief executive officer of Interface Carpet, but everyone, including plant workers, refers to him as "Ray," as though he had just been by for dinner the night before. And now, in an executive suite twenty floors over Atlanta, this unassuming man draws a rough sketch: a circle surrounded by a box, with three arrows pointing into the circle and three arrows pointing out of the circle. He labels the circle "economy" and the box "environment." I take the paper, fold it, and tuck it between the leaves of my notebook, not realizing until later that Ray has just diagramed the philosophy that drives his business practices. Put simply, Ray believes that the economy operates within and is wholly dependent on a healthy environment, but that a strong economy is a prerequisite to that healthy environment. Undeniably, this constitutes unusual thinking for a sixty-eight-year-old man who built a billion-dollar-a-year multinational carpet and textile firm from his life savings and borrowed money.

By his own reckoning, Ray has had three lives. He was an employee of another textile firm until he was thirty-eight years old, then he was an entrepreneur until he reached his sixties, and now he is an environmentalist. But that is not quite right. He is both an environmentalist and an entrepreneur. He is an "enviropreneur," gray haired and neatly dressed, conservative in appearance, calm in demeanor, deliberate in speech, and absolutely radical in thinking. As founder and chairman of Atlanta-based Interface Carpet, Ray has been called "the greenest chief executive in America." Descriptors like "convert" and "zealot" might also fit the man who matter-of-factly states that Interface will become the first environmentally sustainable company in the world.

His own employees refer to him as "a visionary" and talk casually about his environmental "awakening" in 1995. This was the same year that Ray's company and his suppliers, by his own accounting, were responsible for the extraction and processing of one and a quarter billion pounds of material from the earth, of which eight hundred million pounds were burnt. In Ray's own words, this realization made him "want to throw up." He later wrote of his own "haunting role" in the earth's devastation. He called himself, in print, a "legal thief," then went on to say, "My company's technologies and those of every other company I know of anywhere, in their present forms, are plundering the earth."

But this was not enough. Ray went even further: "Someday," he wrote, "people like me may be put in jail."

Ray considers himself a product of both the Great Depression and World War II. In 1973, approaching forty years of age and already a well-established executive with what most people would consider a very comfortable life and a promising future, Ray became aware of a European product that he believed would be welcomed by the American markets. That product was free-lay carpet tiles—eighteen-inch squares of carpet that fit snugly together and that could be used instead of single-piece broadloom carpets that have to be custom cut for every room. Ray saw an opportunity to become the first distributor and manufacturer of free-lay carpet tiles in the United States. His employer did not share this vision, and, as sometimes happens, he and his employer parted ways.

To start Interface, he raided his life savings and borrowed money from friends. "The risk was so frightening," Ray later reported, "like stepping off a cliff in the dark and not knowing whether your foot would land on solid ground or thin air." During startup, he faced the threat of a lawsuit from his previous employer and one of his financial backers dropped out. At one stage Ray claims that he fell to the floor in anguish. But in its first full year of business, Interface made a profit. Within five years, Ray says, he could sleep at night. Ultimately, he became the largest producer of free-lay carpet tiles in the world, making products at twenty-nine sites, selling them in 110 countries, and capturing over 40 percent of the market.

In 1994, just before his awakening, he acknowledges that his company had very little environmental awareness. What little awareness that existed was focused on indoor air quality and compliance with environmental regulations. But then customers—especially customers with design and architectural firms—started asking about Interface's environmental performance. In response, Interface's research people started looking into the company's environmental practices. Ray himself read Paul Hawken's *Ecology of Commerce*, and it changed his life.

Ray was sixty years old when he established a task force that would help Interface become the first environmentally sustainable company in the world. He started the climb up what he and his followers call "Mount Sustainability." But there was one catch: They had to maintain profitability while they climbed. Ray's environmental awakening would not completely smother his sense of fiscal reality. Awake or not, the nerve to the hip pocket remained remarkably sensitive.

Outside of an Interface tufting plant in West Point, Georgia, a sign, posted above well-groomed grass, says, "No Tobacco Products on Property." Inside of the gray concrete building, where I expect dust and clutter and clattering looms, there are spotless floors and neat rows of machinery and a hum of activity too low to hinder conversation. The tufting process converts spools of yarn into strips of carpet, 2 yards wide and 250 yards long. The strips are later cut into squares of free-lay carpet tile. Five days a week, 330 people work in three shifts, producing 50,000 square yards of carpet each day. Electric forklifts, each fitted with its own computer, move about the shop floor. People smile. They chat as they work. This is not Norma Rae's sweatshop.

Joey Milford, a manufacturing engineer, wears an Interface photo-identification badge, showing his gray hair and neatly trimmed gray beard. Attached to the badge is a plastic card with Interface's environmental policy, 114 words long, reminding him to use "less natural resources," to avoid disturbing the "ecocycle," and to "support the local community in developing and maintaining a sustainable environment." Everyone working for Interface has the policy close at hand. The plastic card also defines sustainability: "Meeting our needs in a way that protects the ability of future generations to meet their needs." The card definitely does not say to make less carpet or to slow production or to abandon consumerism. Instead, between the lines, it extols a philosophy that recognizes the link between economic performance and environmental performance.

After fifteen years with Interface, Joey is knowledgeable about every process that occurs here. He also has firsthand knowledge of the history of Interface. He recalls Ray Anderson's environmental awakening—and refers to it as exactly that, "Ray Anderson's environmental awakening." He says that the initial announcement about Interface's environmental mission came at a year-end meeting, when Anderson described his goal of sustainability.

"It hit me as, wow, that's really big for a textile-based industry," Joey says. "Can we really do this? Can we take a textile worker and put these philosophies in front of him and expect to get anything from it?" A first response from many workers was skepticism. While people understood that sustainability made sense, they also believed that costs would overwhelm Anderson's vision.

"We started with the simple things," Joey says. "We started by putting the right trash in the right receptacles for recycling, then evolved to look at how we make our products, what we put in our products, to even taking our concerns to our suppliers and our customers." Joey tells me that Ray's goal calls for reaching sustainability by 2020, but no one can say exactly what that will look like. What is in place is an evolutionary process that encourages integration of environmental concerns with more traditional business concerns at all levels of the company.

In the plant, Joey shows me row after row of metal frames holding spools of yarn. He calls the frames portable creels. Traditionally, carpet yarn would be wound onto beams—large spools that feed the carpet tufting machines but that would, because of their design, leave as much as sixty pounds of scrap yarn behind after each run. Sixty pounds of waste, in Ray Anderson's awakened company, was sixty pounds too much.

By replacing the beams with portable creels, the waste was eliminated.

Would this redesign have happened without the sustainability program? Joey says it would not. It was the sustainability program's call for waste reduction that inspired the new design and encouraged development of the portable creels even though the process required more labor and an investment in new equipment. Nevertheless, the portable creels ultimately paid their own way. The redesign was good for business and good for the environment. As Joey puts it, "The portable creels result in big savings." And this is no exaggeration: New yarn coming into the plant is purchased at something like four dollars a pound, while waste yarn is sold as stuffing for something like ten cents a pound. My nine-year-old son can easily grasp the math. Joey believes that savings from the portable creels amount to more than one million dollars each year. The nerve to the hip pocket is soothed.

In a theme that repeats itself again and again at Interface, it was the sustainability program that allowed employees to explore new ways of thinking that not only protect the environment but also lead to increased profits. It has been said that redesign requires people not only to have new ideas but also to abandon old ones, and that is exactly what the sustainability program encourages.

"Can you make a textile worker an environmentalist?" Joey asks. Every Interface associate—all of the employees are referred to as associates—gets a copy of Ray's book, *MidCourse Correction*, which describes Ray's awakening and extols a new paradigm for business. Some of them even read the book.

"They don't all go home and talk about the environment around the dinner table," Joey says, "but some do. We had twenty employees working on an Earth Day cleanup. What we do here makes people think about what they do at home."

"Our reward for doing the right thing is savings," Joey says. And here he is not just talking about yarn. They hope to reduce the amount of edge that has to be trimmed from new carpet to zero. They are reducing excessive use of flame retardant additives. They are reducing energy use. They are, as they sometimes like to say, "pioneering a better way to bigger profit." *Natural Capitalism* had described waste as money being thrown away, and Sean Skaling at Green Star had agreed. And here, the people of Interface do not like to throw money away.

Part of the formula is efficiency. Inefficiency equals waste—wasted time, wasted energy, and, inevitably, wasted material. With this in mind, one target for Interface is increasing what Joey casually refers to as "first-pass yield."

"First-pass yield," he says, "is about doing this better, with fewer rejects." This is, in part, quality control. But it is also about redesign. For example, a carpet tile that Interface calls "Entropy," inspired by natural patterns such as leaves on the ground and scattered stones, can be laid in any direction; individual tiles do not have to be turned to match the adjacent tile and slight imperfections are invisible.

"Off quality," Joey says, "is low."

When Joey gives plant tours, as he often does, prospective customers seem intrigued by Interface's environmental position. "At first," he says, "it comes to them as a surprise that a textile business down here in southwest Georgia wants to be the world's first sustainable company. But people are starting to see past the Bubba image."

When Joey walks through the plant, he waves at almost everyone. He seems to know all 330 employees by name. He seems to know their families. We walk past portable creels and wave. We walk past stockpiles of yarn and wave. We walk past a room-sized drier and wave. Eventually, we find our way to a prominently placed display. Colorful charts plot progress toward specific environmental goals, tracking such things as carpet scrap, energy use, solid waste, and recycling, with most of the measurements based on yards of carpet produced to account for fluctuating plant production. Environmental performance, like other aspects of business, has to be measured to be managed effectively. The charts grew from ISO 14001 certification, an environmental certification from the International Organization for Standardization; tellingly, and in what many would consider an unusual move, Interface sought ISO 14001 environmental certification before seeking ISO 9001 certification, which encourages and recognizes manufacturing quality. This is a point that Joey presents with measured pride.

"Even with an eighty-thousand-square-foot increase in plant size," Joey says, "we did not increase energy use."

Another mark of pride for Interface, although it does not show up on this particular set of charts, is embodied in an incremental sales volume increase of two hundred million dollars that required neither increased use of raw materials nor increased release of anything that would harm the planet or its inhabitants. Again, increased profits and environmental stewardship walked hand in hand at the Interface plant in West Point, Georgia.

But it is not all good news. I point out that the chart tracking carpet scrap showed lower than normal performance in July. "July was a bad month," Joey says. "Maybe we introduced a new product. Bringing new products on line always causes problems."

The trail that leads to the top of Mount Sustainability has other rough spots. In a corner of the plant, black plastic pallets stand one atop another, stacked well above our heads, a sustainability faux pas. Joey tells me that they bought the plastic pallets to replace wooden pallets. The plastic pallets could be reused almost indefinitely, while the wooden pallets had a more limited lifetime. Also, the plastic pallets did away with the need for disposable plastic shrink-wrap, a requirement for wooden pallets. But plastic burns differently than wood. It is harder to extinguish and its fumes are more dangerous. Insurers said the pallets could not be used. They shut down what seemed like a good move, in terms of both profit and sustainability. Waste is money being thrown away, and for now the pallets are not thrown away; they are stacked in a corner, apparently waiting to be used, hiding from the landfill, a speed bump on the road to the top of Mount Sustainability.

Joey's boss, Ray Anderson, sometimes talks of what he calls "God's currency" and a need to develop what he calls "ecometrics"—a means to objectively measure environmental performance. Ray's language has rubbed off on Joey. It is easy to envision Joey walking through the plant, traversing these well-swept floors, past spools of yarn and portable creels and waving employees, stopping to chat languidly about God's currency and ecometrics.

But for all of this, Joey recognizes that people, as he puts it, "won't buy junk."

"Selling carpet is about look, price, comfort, and durability," he says. "By itself, environment will not sell carpet. You have to give the customer what the customer wants, or the customer goes somewhere else." Somewhere else could be any one of the eight to ten companies that Joey considers to be important competitors.

Worn-out carpet can survive for twenty thousand years in landfills. This includes Interface's share of the almost one billion square yards that find their way to landfills every year. "Some local governments are beginning to require carpet recycling," Joey says. Interface is one of several sponsors of CARE—the Carpet America Recovery Effort. Interface has geared up for recycling, ultimately pursuing what they call "cradle to cradle" manufacturing, designing products from the beginning to be reused. In its Evergreen program, Interface takes this a step further, hoping to convert the carpet industry from a manufacturing and sales industry to a service industry; the customer would, in a very real sense, rent the services that carpet provides. Interface would install the carpet, maintain the carpet, and ultimately remove the carpet. The removed carpet would go right back to the manufacturing stream. The customer would rent the services that carpet provides—appearance and comfort would be paid for by the month just as electricity and water and garbage collection are paid for by the month.

But while sustainability may be the driver behind Evergreen, sustainability is not the sales pitch. "You don't need carpet," Interface tells customers, "you need single-source responsibility, interior aesthetics, comfort underfoot, noise control, cleanliness, protection from slip and fall risk, easy maintenance, flexibility." The capital expense of carpet is replaced with "a deductible monthly payment for floor-covering service." The E word—environment—does not enter the sales pitch until later, until after the customer understands that Evergreen is all about working for the customer, about not selling junk.

Nevertheless, Joey is convinced that Interface's environmental stance offers a competitive advantage. Referring to the companies that he considers to be important competitors,

Joey says, "They're going to have to compete with us on environment."

The corners of his eyes crinkle as he smiles.

Not far from the tufting plant in West Point, David Oakey Designs resides in an office building that could pass for an expensive suburban home from the school of Frank Lloyd Wright: white brick, brown wooden siding, lots of glass. At first glance, the landscaping looks in need of a manicure—it is marginally out of control, a little too haphazard for a suburban neighborhood, definitely not office landscaping. The white bricks are obscured by scrubby brush, too big to be shrubs, too small to be trees. A patch of ground in front of the wooden siding supports overgrown grass and weeds, brown this late in the year, with bushy seed heads. But on second glance the impression turns from one of neglect to one of blending. The building, to the extent that it is possible for a building to do so, fits in with the surrounding pinewoods. The landscaping is a transition between the building and the pinewoods. It dawns on me that all of this may be intentional. This is David Oakey, chief artistic designer for Interface floor coverings, expressing himself. This is biomimicry in action, the same technique that David applies to carpet.

"Biomimicry," David tells me, "imitates nature." Bionics, he explains, is a field of engineering that looks to nature for engineering advantages. "Torpedoes designed to mimic tuna," he says, in an accent sounding faintly British, perhaps, but mixed with other places and slowed by deliberate thought, or perhaps slowed by association with his neighbors here in the southern United States. From bionics to biomimicry, he seems to be saying, is a natural step forward. Biomimicry lets him spot the appeal of chaos in nature, and this chaos, applied to carpet tiles, both pleases the eye and the bottom line. Drying leaves scattered on the ground, pebbles in a streambed, clouds in the sky—these are inspirations for the Entropy line of carpet tiles that, as Joey Milford pointed out, increase first-pass yield and decrease off quality.

David has a picture of a hotel room's carpet branded with the outline of a clothes iron. "What do people do in hotel rooms?" he asks incredulously. But more to the point, he suggests what hotel operators can do with carpet tiles, especially those like the Entropy line that mimics the chaos of nature: The burned carpet tiles can be replaced without shipping the entire roomful of carpet to the landfill. Even if the surrounding carpet is slightly worn or faded, the difference between it and a new carpet tile, with a natural pattern, will go unnoticed. A chambermaid can replace a damaged tile with a new tile from her cart, or the damaged tile can be switched for a good tile from under the bed. Ideally, the damaged tile with the burned outline of an iron will be recycled. For the hotel operator, there is no downtime for a room and there are no labor charges for a carpet crew. For the world, a roomful of carpet stays out of a landfill and neither energy nor material are wasted. Efficiency increases and waste decreases.

"When Ray Anderson started this dream of a sustainable company," David Oakey says, "I didn't believe him. I thought he was green-washing. That's the phrase he uses—green-washing. It was hard for me to picture what a sustainable company would look like. I thought of wool and cotton and hemp and I knew that wouldn't happen. I knew the public wouldn't accept paying higher prices."

After reflecting for a moment, he says, "Ray was talking about sustainability, but I didn't even know what it would look like."

"Do you now?" I ask him. "It seems to me that this is an evolutionary process."

"Evolution. That's what it is. I have no idea what it will look like in the end. What will our products look like in the end? How will we make them? Sometimes as you go along in what you thought was the most environmental way to move, you suddenly find a roadblock."

As an example, a supplement to the stacked plastic pallets at the West Point plant, he points to recycling. He was surprised when he discovered that certain kinds of recycling required more energy than using new material. "The products and processes were engineered from our mass-production industrial age, just making things faster and cheaper. Now, how do you change that whole process? The product should work for the customer, but it should come back for recycling. Maybe a new polymer that can be easily recycled is what's needed."

His thinking goes beyond carpet and beyond Interface. He talks of the need for cooperation between companies as an environmental necessity. "I go back to nature as my inspiration," he says. "I don't think a single tree can be sustainable, but a forest or an ecosystem can be sustainable." He talks about competition and cooperation. Because of competition, other carpet manufacturers are starting to follow Interface's environmental lead. Interface cooperates by sharing information. But companies can cooperate in other ways. One company's waste can be another company's raw material. In cogeneration, one company can generate electricity, and the waste heat from the generator can be piped to another company to heat offices or to run driers. In trading carbon credits, one company can reduce emissions, then sell the credits for the reduced emissions to another company. Ultimately, competition can be complemented by a network of cooperation, companies helping one another in a web of enlightened self-interest.

David likes cars. But he does not like ugly cars. David does not like ugly cars even if they run on modern environmentally friendly technologies, like batteries and fuel cells. Hybrid cars, to David, are ugly. "Don't buy it because it's a green car," he says. "Buy it because it's a fun car. Why can't good design be green design?"

He drives the car analogy further. "I'll relate Interface to the automobile industry," he tells me. "I'm going to make an automobile that will get me from A to B using the least amount of gasoline possible. Can I make a carpet using the least amount of materials, the least amount of fossil fuels, to make the product aesthetically pleasing and with the performance expected by our customers? We reduced material use in carpet and went from a twenty-eight-ounce face weight to just over twenty ounces."

He calls this "dematerialization"—the reduction of material used to make a product without degrading the quality of the product. The idea is simple: To the extent that a product is overbuilt, waste is incorporated into the product. Interface researchers discovered that they could reduce the amount of yarn used in carpet from twenty-eight ounces to twenty-two ounces per square yard without a change in appearance or durability.

In a slide show David is preparing for an Interface annual meeting, the word "waste" covers the screen, and then, in computerized animation, the message becomes, "Waste is lost profit."

"The waste factor still can be our biggest gain on sustainability," he says. "We just waste too much stuff, whether it's gasoline, carpet, food, anything. In this country, we're just not conscious of what we throw away. And for business it's lost profit."

In the foyer near the building's exit, a booth sells purses. The purses are both attractive and pricey. A spin-off company called Neodesign makes and sells the purses, using carpet scrap as the raw material. The waste from David's shop has become expensive Neodesign purses. On a per yard basis, the purses cost hundreds of times more than carpet and probably thousands of times more than carpet waste.

"People won't buy junk," Joey Milford had told me, and David has asked, "Why can't green design be good design?"

I walk out, past the overgrown and browning grasses and weeds, under the shade of the scrubby brush. No one is wasting time cutting the grass or trimming the shrubs. Green Star does not operate in Georgia, but if they did, Interface would be their poster child.

In less than four years, Interface reduced total waste by 40 percent, saving the company sixty-seven million dollars a year and improving the bottom line by about 7 percent. This is why Dan Hendrix, president of Interface, Inc., can sit in a conference room high above Atlanta and, along with other Interface executives, pepper his conversations with phrases like "intergenerational equity," "renewable resources," and "environmental performance." This is why they can use the phrase "climbing Mount Sustainability"—a phrase that seems to roll off the tongues of almost everyone at Interface—without even a hint of a blush or a smile. This is why they can tell me that they hope to see Interface "disconnected from the wellhead." This is also why Mike Bertolucci, head of Interface Research Corporation, can spend as much as 80 percent of his budget on sustainability issues.

I mention that I had visited the plant in West Point and been impressed by the interest of the plant employees in the environmental aspects of Interface. "At any one of our locations," Mike says, "you'll find that Interface employees take real pride in the company and its sustainability program. The way we have integrated the entire process, right down to the floor, is probably one of the most remarkable aspects of the cultural change here at Interface."

"It's definitely become part of our culture," Dan adds. "Within the company, we can talk about it naturally. We don't need a forced agenda."

I suggest that what they have done at Interface is comparable to what has happened with safety practices over the past hundred years. In the early 1900s, safety statistics in manufacturing, construction, and mining were abysmal; death, dismemberment, and trauma were seen as an acceptable cost of doing business. Over time, this became less acceptable. Business transitioned. Safety officers became commonplace. Every employee participated in safety meetings. A cultural shift occurred. Today, accidents and even incidents—the close calls that could have hurt someone but did not—spark investigations. No one can imagine otherwise. Businesses track lost-time accidents and days away from work just as they track other aspects of performance—and just as Interface tracks environmental performance on its ISO 14001 charts in the West Point plant.

"To me," Mike says, "it's more like the quality movement of the eighties. If people aren't doing this in the twenty-first century, they're not going to be in business. It's just going to become a requirement of doing business."

"It's more than just sustainability," Dan says. "Sustainability is one leg of a three-legged stool. The other two legs are social responsibility and environmental soundness. You'll get a perspective of how these things interplay by looking at our sustainability reports." Dan's voice is even and quietly reassuring throughout; he communicates as a business leader, self-confidently presenting as routine what are in fact revolutionary ideas. His tone and his body language say, "This should make perfect sense to everyone."

"Every one of our employees does not think of it this way," Mike says. "A high percentage of them do, but not all of them. Lots of our employees probably think of it in terms of recycle content, which is a big part of our story, but it's not the whole story. They talk about renewable materials and renewable energy, but when we talk about Mount Sustainability, it's continuing education for us."

"The challenge in the marketplace," Mike says, "is really defining what sustainability means." One thing it means is reducing fossil fuel use, or what Interface associates sometimes call "carbon intensity." From 1995 to 2001, Interface reduced use of fossil fuels by more than 30 percent. Another thing it means is, as Interface associates sometimes put it, "getting ahead of environmental regulations." Another thing it means is understanding and managing life-cycle costs, recognizing, for example, the cost to the environment from customers who throw away used carpet long after the conventional accounting books have been closed. It means, too, recognizing and avoiding the so-called perverse subsidies, or the kind of incentives that encourage wasteful practices. It means, for example, ending tax advantages or government subsidies that support traditional practices even though they may not be good for the environment, such as those enjoyed by the automobile industry, which, according to some authorities, reaps the benefits of more than a trillion dollars in what is sometimes called corporate welfare. It means dematerializing everything, using less material to provide the same service, as in reducing carpet face weights from twenty-eight ounces to twenty-two ounces. It means thinking about the links between various aspects of the manufacturing process, understanding, for example, that a 4 percent reduction in nylon results in a 200 percent energy savings.

"The Evergreen Lease concept is maybe fifteen years ahead of its time," Dan says. The Evergreen Lease is a key part of sustainability in that it will ultimately ensure that carpet sent to the customer will return to the plant for recycling and reuse. But while programs like the Evergreen Lease may sound very sensible, they are not necessarily easy to sell.

"Almost none of our business is Evergreen Lease," Dan says. "Part of the issue is that the customer's budget has to be in one place. Usually the janitorial budget is over here, the capital improvements budget is over there, the maintenance budget is somewhere else. For Evergreen to make sense, those budgets need to be seen together. And the IRS guidelines won't let you treat carpet as an operating lease. You have to capitalize it. So the IRS gets in the way."

Mike chimes in. "There's an alignment today between the first industrial revolution concept of 'take, make, and, waste' from a materials standpoint and 'buy it, depreciate it, and throw it away' from an economics standpoint," he says. "The concepts are absolutely parallel. We see the same thing with sustainable business and the Evergreen Lease concept. The Interface view of the sustainable model for business is cyclical in its processes. It's like nature, where waste equals food, where one organism's waste is another's food. In the Evergreen Lease concept, you never release the material content of what it is you're selling, you only release the services component. Sustainable business and the Evergreen Lease concept are in alignment. They're parallel. You put these two concepts together and you have what we believe is sustainable design."

"But with what Mike just explained," Dan says with a chuckle, "you've just lost most chief financial officers."

These are businessmen talking. These are businessmen who run a very competitive business and run it successfully—businessmen who commute into work every morning, who wear suits and ties, who understand stock reports and balance sheets. They just sound like granola-chewing environmentalists.

"When Ray Anderson first started talking about his awakening," Dan tells me, "Wall Street thought he was terminally ill. I had an investor call me and say, 'You know, I followed Interface for twelve years, and something is wrong with Ray.' And people started looking sideways at him."

Four years later, Interface revenues had doubled, employment had doubled, and profits had tripled.

Ray Anderson is reluctant to exploit green practices in advertising. "We don't know how to do it without risking the greenwash label," he says, but he remains confident that green practices attract clients. He is so confident, in fact, that he approved installation of solar panels at an Interface plant in California despite the high cost of solar electricity. When his accountants warned him that the panels would be a bad investment, he explained to them that higher costs for electricity would be recuperated through increased sales, picking up that slight but meaningful tinge of competitive advantage that the environment offers. He also believes that sustainable practices have helped stock prices, have helped attract and retain good employees, and have helped relationships with regulatory agencies.

Looking into the future, Ray believes that markets as we know them today will not survive if businesses refuse to develop sustainable environmental practices. "The prototypical company of the twenty-first century," he says, "will be taking nothing from the earth, emitting nothing, and wasting nothing." A time will come, Anderson believes, when measures he is putting in place now will become mandated by legislation. At that time Interface, with programs already in place, with costs internalized, will have a tremendous business advantage.

Anderson's vision extends beyond his own company's practices to those of his suppliers. Interface espouses pursuit of what they cryptically refer to as "sensitivity hookup," which calls for "creating a community within and around Interface that understands the functioning of natural systems and our impact on them." And this community clearly includes Interface's supply chain. Anderson once pointed out that Interface comprises less than one-thirty-thousandth of the global economy, while Dupont, one of his company's key suppliers, comprises one-six-hundredth of the world economy. "You are your supply chain," he says, and with that philosophy in mind he pressures his suppliers, including giants like Dupont, to stop and think about their own processes and practices.

Anderson recognizes realities that most people ignore. For example, he knows that a ten-pound laptop computer results in forty thousand pounds of waste, that for every laptop computer owned by Interface—and, of course, Interface is pragmatic enough to recognize the value of laptop computers—there exists somewhere a small mountain of mine spoil and a stream of effluent. He knows that the amount of land needed to support the lifestyles of Americans exceeds the size of their nation by 20 percent. He knows that something like four million pounds of raw and processed material is moved for every typical American family every year. Yet he by no means espouses asceticism. He would agree with the World Bank's statement saying, "The key is not to produce less, but to produce differently." He recognizes the codependence of economic growth and environmental protection. In his sketch, arrows pointed both ways to join economy and environment.

"Education is power, and the public tends to be ecologically ignorant," Anderson laments. To address ignorance, he talks publicly and donates money for environmental education. He sat on the President's Council on Sustainable Development. His book about sustainable business practices has been printed in English, Japanese, and Chinese, and he recently helped dedicate the School of Environmental Science at Beijing University. Most importantly, he educates through his actions.

His support of sustainability might be less noticeable if his company manufactured products more often associated with the environmental movement, such as wind generators or recycled paper or park benches reformed from used plastic soda bottles. But when a leading manufacturer of carpets and textiles moves his company toward sustainability, and increases profits in the process, how can it be ignored?

"The word is getting around," Anderson says, "that sustainability is a thing to be paying attention to."

CHAPTER NINE

THE REBUTTABLE PRESUMPTION

Resources for the Future, a nonprofit organization formed during the Truman administration, has its own building fifteen blocks north of the White House. In Truman's time, people worried that the world was running out of basic resources, things like aluminum and iron and oil. In its early days, scholars at Resources for the Future showed that advances in technology and continued discovery of new materials would more than compensate for resource scarcity, but they recognized that ever increasing use of resources would come with increasing environmental costs. While there might be enough aluminum and iron and oil, there might not be enough environment. Resources for the Future became a leading environmental economics think tank, perhaps *the* leading environmental economics think tank, describing itself as a place where "scrupulous academic research intersects with acute policy relevance." This seemed to me a good place to look for the big picture view, a place that might understand in broad terms how the things I had seen and heard about—the flow of money in the name of the environment and what that money bought—could coalesce into a simple but meaningful message.

Len Shabman, once an academic, now works here as a resident scholar. He has made a career studying the economy and the environment. As an undergraduate more than three decades ago, well before the concept became fashionable, he wrote a thesis espousing environmental taxes to support water pollution control. Early in his career, he testified against the Tennessee Valley Authority, discrediting an economic analysis supporting the construction of a dam. As a professor, he took leaves of absence to work within the federal government. "A good share of my time as an academic," he says, "was devoted to trying to make decisions—regulatory decisions, budgetary decisions. I've written a lot of regulations. I've made thumbs-up thumbs-down budget decisions. To the extent that I have any views on how economic thinking interplays with environmental issues, they come out of those experiences rather than from writing academic papers."

I show him a copy of Grant Ferrier's *Environmental Business Journal* with its bar graph plotting environmental industry revenues in the United States. The bars climb quickly through the 1970s and into the 1980s, but then plateau at $230 billion a year. I ask if he thinks $230 billion is a lot of money.

He coughs. He adjusts his glasses. He does not answer.

I rephrase the question. "Does the number surprise you?" I ask.

"My first reaction," he says, "is no. It seems low to me. But I don't know what it is in reference to. If I did a similar survey, where would I draw the boundaries? What would I include? Two hundred and thirty billion doesn't surprise me at all."

I ask if he thinks we might be overspending. "We certainly haven't overshot the mark on total spending on the environment," he says. This leads me to mention a widely circulated estimate claiming that the environment provides thirty-three trillion dollars worth of services each year. Len remembers that a former colleague once quipped that thirty-three trillion dollars seemed to him a gross underestimate of infinity.

I read aloud some of Grant Ferrier's statements about money and the environment: "What is needed now is economic consequence" and "Economics should be considered above all else" and "It should all be monetized." And there is Grant's faith in "an assessed lifecycle economic value of the land impacted" and "an incremental negative economic consequence to each increment of unsustainable behavior." It seems to me that Len would, on the face of things, agree with statements like these. But when I relay to him Grant's belief that the way forward is through a better understanding of the value of our environment, through putting dollar values on the environment, then using these values to set up a pollution tax system—a user-pays approach to protecting the environment—his response is unequivocal.

"I strongly disagree," he says. "There are many economists who would agree with you, but I am not one of them. I think we can make sound environmental decisions without these numbers."

After seeing water treated in New Orleans, after watching the army work through an environmental impact statement for its Stryker transformation, after driving up Mount OII and hearing Ray Anderson talk about climbing Mount Sustainability, after learning what it takes to breathe clean air—after all of this, one thing seemed clear. People are struggling with tough choices about how to spend what is, in the end, never enough money: a billion dollars for water system repairs and upgrades in New

Orleans; a million dollars for one environmental impact statement, with something like two prepared each day since 1979; almost a billion dollars for Mount OII, one small piece of land in southern California; over a hundred billion dollars in estimated costs for clean up of all of our Superfund sites; six billion dollars a year spent by nonprofit environmental organizations; billions more to replumb the Florida Everglades; eighty billion dollars and counting to make the nation's rivers and lakes swimmable and fishable.

"A former professor of mine," Len says, "used to say that anything worth doing is worth doing inefficiently. And a lot of our environmental regulations might be described that way. But today we know better. We know how to regulate smarter. As we move forward, we can do it more efficiently. Then we can have more environmental quality and more of other things, too."

He seems to see money not as something to have and to hold but as "a measure of choice." In a fair market, he believes people will spend their money on those things they value most. But this does not work if the market is slanted.

"Money is no object if it isn't your money," he says. "Too many of our policies have everybody paying for everybody else's environmental quality. That's almost a guarantee of inefficient decisions. That really concerns me. It concerns me that there is no consideration of costs." Here, I cannot tell if he means that environmental laws are applied without considering the costs, or if he means that a developer or a manufacturer can degrade the environment without considering costs. My feeling is that he means—whether he intends to or not—both.

"The market is not engaged," Grant Ferrier had told me, "and investment dollars will not shift unless there is a consistent economic consequence to unsustainable behavior." In this regard, Grant Ferrier and Len seem to share a common view of the world.

Pointing to the stagnant spending on the environment, hovering at $230 billion a year, Grant had talked about a sense of dissatisfaction in the environmental movement. Stagnation could signal that environmentalism has succeeded and that more money is not needed, but no one who moves through life with open eyes can believe this. It could signal that the economy is in a slump and that money for the environment is just not available, but the fact that environmental spending as a portion of the gross domestic product has shrunk belies this possibility. And it could signal that people have lost interest and moved onto other things.

Len points me to a series of articles run by *Grist*, an online environmental magazine. An Internet discussion was sparked by an essay written by Michael Shellenberger and Ted Nordhaus; they delivered the essay, which they called "The Death of Environmentalism," at a meeting of the Environmental Grant-makers Association. A quote: "Our parents and elders experienced something during the 1960s and '70s that today seems like a dream: The passage of a series of powerful environmental laws too numerous to list, from the Endangered Species Act to the Clean Air and Clean Water Acts to the National Environmental Policy Act."

From Peter Teague, environment program director of the Nathan Cummings Foundation: "The environmental community can claim a great deal of credit for what are significant advances over a relatively short period—advances won against well-financed campaigns of disinformation and denial."

From a survey of fifteen hundred Americans by the public relations firm Environics: Respondents who agree with the statement, "To preserve people's jobs in this country, we must accept higher levels of pollution in the future," went from 17 percent in 1996 to 26 percent in 2000. Respondents who agreed with the statement, "Most of the people actively involved in environmental groups are extremists, not reasonable people," went from 32 percent in 1996 to 41 percent in 2000.

From Susan Clark, executive director of the Columbia Foundation: "The problem is not external to us; it's us. It's a human problem having to do with how we organize our society. This old way of thinking isn't anyone's fault, but it is all of our responsibility to change."

From Van Jones, a proponent of an environmental movement that works closely with labor unions, civil rights groups, and businesses: "The first wave of environmentalism was framed around conservation and the second around regulation. We believe the third wave will be framed around investment."

From an anonymous author, commenting on "The Death of Environmentalism" for *Grist* magazine: "The elephant in the room is funding. Yes, money—how it's spent, where it comes from, and what is demanded in exchange. Hundreds of millions of dollars a year are poured into environmental causes by foundations, corporations, and individual donors. That's a lot of dough-re-mi."

More from the authors of "The Death of Environmentalism," Michael Shellenberger and Ted Nordhaus: "Environmentalism is today more about protecting a supposed 'thing'—'the environment'—than advancing the worldview articulated by Sierra Club founder John Muir, who nearly a century ago observed, 'When we try to pick out anything by itself, we find it hitched to everything else in the Universe.'"

Shellenberger and Nordhaus do not say it, but the one thing to which the environment is certainly hitched is the economy.

Len tells a story about a discussion in which he was arguing that something—some environmental program—was more costly than necessary. The person he argued with said, "No, we have these standards and rules. The cost be damned."

"And I suddenly realized," Len tells me, "this principle of the rebuttable presumption. Ever since then I've seen the world differently."

He leans forward in his chair, his elbows on his desk and his hands clasped in front of him. "Years ago," he says, "environmental improvements had to be justified starting from the current condition of the environment. The burden was on the proponents of an improved environment. Numbers were thrown around. People would ask, 'Are the benefits bigger than the costs? Is additional environmental spending and regulation justified from a cost perspective?' But when it comes to deciding about what to do for coastal Louisiana and the Everglades and the Great Lakes and whether you should take out the last microgram of PCB from the Potomac River, I don't think we are going to answer those questions entirely with that kind of economics. We have come to a point where the rebuttable presumption is that you are going to take the last microgram out. The rebuttable presumption on a wetland is that you will avoid filling it. The rebuttable presumption on a natural resource damage-assessment case is that you are going to put it back like it was. That's the starting point of the argument. It is not, 'How does it look today and can we justify making it better?' The starting point is some ideal. The starting point is, 'How should it look?' Then the economic argument comes in and you try to challenge people. You confront them with the possibility that a little bit less than the ideal would save a lot of cost and might be acceptable. You point out that in tough spending times you can get closer to the ideal if policies are more cost efficient."

In Len's view, the shift to what he calls the rebuttable presumption is a key environmental movement victory—maybe *the* key environmental victory. The Clean Air Act, the Clean Water Act, the Endangered Species Act, the Resource Conservation and Recovery Act, Superfund—all of these are results that emerged from shifting expectations. The starting point is not the reality of a trashed planet burdened with too many people and too many careless acts. The starting point is an ideal.

Several years ago, Len and a coauthor wrote an academic article called "Environmental Valuation and Its Economic Critics." The first sentence read, "In the last few decades economists have devoted significant professional attention to developing and applying methods to place monetary values on environmental services." Economists have used surveys and market analyses to try to understand what people think clean water is worth, or clean air, or a bird, then contributed these values—expressed as dollar values—to policy debates. They are asking, in a sense, how much people are willing to pay for clean water and air and living birds. A snippet from Len's article: "Economic valuation, at a conceptual level, is said to be a measure of the preferences people hold for different states of the environment." Another: "The critics all question one, or both, of two core assumptions: that choices made in real or hypothetical markets can be interpreted as a reflection of preferences or value; and that such interpretations should direct decision making." They quote other academic authors. One says, "To believe that markets determine value is to believe that milk comes from plastic bottles." Another asks, "Why should we regard the satisfaction of preferences that are addictive, boorish, criminal, deceived, external to the individual, foolish, grotesque, harmful, ignorant, jealous, . . . or zany to be a good thing in itself?"

For this paper, Len and his coauthor won an award from the editors. The editors considered it the best paper of the year. "We wrote it," Len tells me, "to say, 'Don't get snowed by economists who tell you they know how to put money values on the environment. Everyone is not on the same page about this stuff.'"

Maybe this is where Len differs from Grant Ferrier. Grant seems ready to assign values—and he would assign very high values—to the environment. Under Grant's system, users would pay those values whenever they harm the environment, and damage to the environment would wane.

"Some think they can figure out what a beautiful sunset is worth," Len says, "what the existence of an endangered bird is worth, and so forth. They think they can put a monetary value on the environment and, based on that monetary value, decide whether to put more smoke in the air or harvest the last whale. But when real money is at stake people will argue endlessly over the calculated values. They'll argue over whether a bit more oil production is worth more than a bit smaller wildlife area. It is kind of a value arms race. My number is bigger than your number. And I've always felt that this gets us nowhere. Some things to me philosophically don't lend themselves to monetization. It's like asking what it is worth to have the right to vote. It's a monetization of something that people don't want to see monetized. I think this is a distraction from a necessary public debate, a debate we should have without the fog created by these value numbers."

But Len is far from ready to give up on monetization altogether. "I want to do monetization, but I want to do monetization that is going to create more light than heat. If I'm going to do monetization, I want to do something credible. You can't monetize everything and come up with a bottom line benefit-minus-cost number that tells you the right answer. That's why we have public discourse and politics. Some things can be monetized and some things can't. Here's an example. If I know that building a certain kind of marsh in Louisiana will slow down a storm surge, and that's going to have a certain kind of effect on the wave energy, I think it's entirely plausible to monetize that service. We're putting a monetary value on the marsh not because it is beautiful or because it might support endangered species but because it has utility. It creates an economic service." The marshes he describes surround New Orleans and Houma and Morgan City, and the tall grass and reeds of these marshes absorb wave energy. Although the marshes have been greatly diminished over the past few decades, without what is left of them Hurricane Katrina would have been even more devastating. Without these marshes, the cities they protect would have to be abandoned to the ravages of hurricane waves from the Gulf of Mexico. The ability of these marshes to protect roads and houses and pipelines has a value, just as a rock breakwater or steel sheet piling would have a value.

"With this kind of monetization," Len says, "we can better manage our budgets. Within Louisiana we can decide which projects are most valuable, and within the nation we can decide if restoring Louisiana is more valuable than, say, restoring the Chesapeake. But where would you draw the line?"

One could monetize everything, but Len sees no usefulness in this. One could monetize those aspects of the environment that provide services with a clear value to the economy, which he sees as somewhat more useful but fraught with challenges. But there is, in Len's view, a third use for monetization. One could set limits—a cap—on how much pollution can be released or how many acres of land might be developed, still allowing for a sustainable environment, an environment that people could embrace. "After all," he says, "the environment can accept, process, and store some of the waste products of the economy. Pollution happens when we overdo it. So once we decide how much is enough, how much we can live with and without, we let people buy and sell the rights to use the environment. We have a market in these rights, and we let the market do what markets do best—promote cost efficiency and reward innovation. In this case the power of the market for cost efficiency and innovation is directed to a better environment."

In Len's vision, the cap-and-trade approach used in the clean air arena could be more broadly applied. "I don't really call that monetization, though," Len says. "I call that creating environmental markets." Power plants trade sulfur dioxide emissions. Developers trade restored wetlands for permission to build on marshes and swamps. "In any group of regulators," Len says, "it's almost a sure thing that someone will say, 'I just can't come to grips with somebody making money by doing something for the environment.' It just doesn't fit with their image of what markets do. I don't know if it's a fundamental philosophy. Mostly, I think they were taught in school about this thing called market failure. They see the market as the problem, not as the solution. To tell them that we are going to bring the market in and their program will be better off, that just doesn't compute."

Len says, "Even industry has trouble with the idea at times. Early on in the sulfur-dioxide trading program, companies were leaving the program to their environmental compliance offices. But when it became clear that there was actually a financial angle to buying and selling allowances and deciding how much to emit and how much to not emit and how much to buy—when that became clear, it all became part of the business line. It moved it from the environmental office to the business office. That's when it took off. As long as it was viewed as a constraint instead of an opportunity, that's the way it is going to be. And one of the reasons it was seen as a constraint is because that is the way we usually design our environmental regulations."

Earlier, Len expressed concern that some environmental regulations—what he and others call "command-and-control" legislation—had led to inefficiencies. "We're only going to make progress if we start doing things like trading and pricing," he says now. "We're only going to achieve our environmental goals in an efficient way through things like trading and pricing. The buzz word I've picked up is 'second-generation environmental law.' It's very much organized around performance incentives rather than command-and-control."

Grant Ferrier had dreamed of changing the world. Everyone I talked to over the past two years shared this dream. Often, they seemed distracted by the trivia of day-to-day survival, of staying within budgets, of running between meetings. Within this, they lived with the frustrations of incomplete successes. But, almost without notice, the dream became reality. The world changed. While they worried about bottom lines and the money spent for what were often incremental gains, they were all part of this change. They all contributed. The excitement of the early 1970s disappeared, but almost without exception our rivers no longer catch fire. We dig up and destroy springs of toxic goop. We stop to think through the environmental implications of mines and highways and dams. Ecologists are no longer kooks. Earth Day is a regular event, marked on some calendars, just like Columbus Day and President's Day and Christmas, and celebrated not only by environmental organizations but in schools and government offices and even in the lobbies of some oil companies. A scientist can talk seriously of "negotiating win-win situations for conservation and economics." For fear of liability, companies avoid producing hazardous waste. A company may burn coal, but it treats the emissions with baghouses and catalysts seldom considered or even imagined thirty years ago. The technology companies behind clean air equipment have their own trade association. People can start profitable companies that build wetlands. While some may once have thought of Lois Gibbs as nothing more than the hysterical mother of a sickly child from Love Canal, she is known now as the mother of Superfund. Construction has been completed at nine hundred Superfund Sites, and almost three hundred sites have been cleaned up and deleted from the Superfund list. At Mount OII, what the government once called "the potential for an explosion or fire" is gone. People pay to recycle computers. Businesses seek out environmental certifications. There is room now for a radically neutral environmental organization that considers itself both pro-environment and probusiness.

Over twenty years, more than a trillion dollars have been spent, and we are not broke. The environmental movement is part of the business world. While there is room to take things further, economic consequence for environmental performance is a reality. Companies cannot function without environmental managers. Industry relies on consulting firms that provide expertise in air and water and chemical contaminants. Environmental activism has itself become a business worth six billion dollars a year.

Len Shabman's rebuttable presumption has changed from the degraded world of the here and now to an ideal and pristine past.

It sometimes seems as if the only real victories in the environmental arena are the new regulations, the court orders to cease and desist, and the permits that do not get approved. But there is victory, too, in efficiency, in cost-effectively reducing or reversing an environmental impact. There is victory in doing more for less, stretching that $230 billion as far as it can go and leaving cash on the table for health care, nice homes, vacations, education, social work, art.

"I used to think people should care about bang for the buck," Len says. "If you had a fixed budget to spend on the environment, you would want to do as much with that budget as you possibly could. Conceptually, people seem to agree, but the argument gets lost. At times it seems like talking about money and the environment in the same breath is off limits. Maybe it's because the environment is too important to talk about in terms of money. Money is a tawdry subject."

Or maybe, in the environmental arena, there are few rewards for efficiency. Cost effectiveness lacks glamour. "Efficiency is not," Len says, "a very good rallying cry."

Somewhere out there, Stuart Paulus runs a public meeting, explaining how a proposed project might affect the environment. Mike Rolband wades through a marsh, his dogs splashing behind. Gordon Austin applies for grants to improve water management in New Orleans. The Thermal Destruction Plant at the base of Mount OII turns toxic leachate into electricity.

Sean Skaling assesses an Anchorage business. Joe Kubsh and Dave Foerter brief government regulators on an emerging technology. Thread spools off of a portable creel at an Interface carpet plant.

Outside Len's office window, across the roofs of Washington houses and offices and restaurants, out past several blocks of cars and buses and trucks leaking greenhouse gases, I watch a crane lifting something to the top of a building. It is the kind of crane often seen on building sites, with a vertical tower going up to a rotating boom. From Len's window, I cannot know what people are doing in their offices, or where they are driving, or what the crane lifts. It is a busy day in busy town, and there is work to be done. Much has been accomplished, it seems, but still there is work to be done.

INDEX